OXFORD REVISION GUIDES

AS & A Level

CHEMISTRY

through diagrams

Michael Lewis

OXFORD

UNIVERSITY PRESS

OXFORD
UNIVERSITY PRESS

Great Clarendon Street, Oxford OX2 6DP

Oxford University Press is a department of the University of Oxford.
It furthers the University's objective of excellence in research, scholarship,
and education by publishing worldwide in

Oxford New York

Auckland Bangkok Buenos Aires Cape Town Chennai
Dar es Salaam Delhi Hong Kong Istanbul Karachi Kolkata
Kuala Lumpur Madrid Melbourne Mexico City Mumbai Nairobi
São Paulo Shanghai Taipei Tokyo Toronto

Oxford is a registered trade mark of Oxford University Press
in the UK and in certain other countries

Database right Oxford University Press (maker)

First published 2000

Second edition 2001

British Library Cataloguing in Publication Data

Data available

ISBN 0-19-914198-3

10 9 8 7 6 5 4 3

Designed and typeset in Optima
by Hardlines, Charlbury, Oxfordshire UK
Printed in the United Kingdom
by goodmanbaylis

CONTENTS

How to use this book

- Be clear exactly which specification you are following.
- Obtain a summary of the main contents of the AS units and the A2 units.
- This book covers AS and A2 material for all specifications and therefore there are some sections that you will not require.
- Note too that some sections of this book will only be wanted by A2 students, but the level required may vary from one specification to another.
- The material in the book is not divided up into AS and A2 because this may vary from one specification to another.
- Some sections are included to address some of the requirements of Options or Special Studies. No attempt has been made, however, to cover all the content of all such optional work.

EDEXCEL – NUFFIELD – SPECIAL STUDIES	
Biochemistry	Protein structure, Protein synthesis
Chemical Engineering	Kinetics, Phases, Phase equilibria, Raoult's law, Chemical industry
Food Science	Protein structure
Materials Science	Polymers I, II, Modifying structures, X-ray diffraction
Mineral Process Chemistry	Raw materials – Formation of ore bodies, Exploitation

OCR – OPTIONS	
Biochemistry	Protein structure, Protein synthesis
Analysis & Detection	Infra-red and NMR spectroscopy, Chromatography, Molecular spectra and X-ray diffraction
Gases, Liquids, and Solids	Phases, Phase equilibria, Raoult's law, Homogeneous mixtures
Transition Elements	d-block elements, Colour
Environmental Chemistry	Oxygen chemistry, Environmental chemistry

Specification structures

	EDEXCEL	EDEXCEL (NUFFIELD)	OCR
A S	AS 8080	AS 8089	
Unit 1	1 h 20 min　　　　AS 30% A 15% Structured questions	1 h 30 min　　　　AS 30% A 15% Structured questions	1 h 15 min　　　　AS 30% A15% Structured + extended questions
	Atomic structure; formulae, equations and moles Structure & bonding; Periodic Table I – intro. Redox intro. Gp I, Gp II, Gp VII	**Introducing Chemistry** Metal compounds; Alcohols; Periodic table and atomic structure Acid–base reactions; alkaline earths Energy change and reactions	**Foundation Chemistry** Atoms, Molecules and Stoichiometry Atomic structure; Bonding and structure Periodic Table: Gp II and Gp VII
Unit 2	1 h 20 min　　　　AS 30% A15% Structured questions	1 h 30 min　　　　AS 40% A 20% Structured questions	1 h 15 min　　　　AS 30% A 15% Structured + extended questions
	Energetics I – Hess' law Organic – alkanes, alkenes, halogenoalkanes, alcohols Kinetics – qualitative Equilibria I – qualitative Industrial inorganic – Haber proc., Contact proc., Aluminium extraction	**Bonding and Reactions** Redox reactions and halogens Covalency and bond breaking Organic – hydrocarbons; Intermolecular forces Organic – halogenoalkanes	**Chains and Rings** Alkanes, alkenes; Fuels Alcohols, halogenoalkanes
Unit 3	Internal Assessment　AS 40% A 20% or Practical Test 1 h 30 min + Data Interpretation 1 h	Internal Assessment　AS 30% A 15%	1 h　　　　AS 20% A 10% Structured questions + Coursework or　　AS 20% A 10% 2 h practical test
	Experimental Skills	**Practical Skills**	**How Far, How Fast** Enthalpy changes Reaction rates Chemical equilibrium + **Experimental Skills**
A2	A 9080	A 9089	
Unit 4	1 h 30 min　　　　A 15% Structured questions	1 h 30 min　　　　A 15% Structured questions	1 h 30 min　　　　A 15% Structured questions
	Energetics II – lattice enthalpies Period. Table II – Per. 3 and Gp 4 Equilibria II – quantitative Acid–base equilibria Organ. chem. II – acids, esters carbonyls, acid chlorides, nitrogen compounds, further halogenoalkanes	**Energy and Reactions** Rates of reaction; Arenes – benzene and phenol; Entropy; Equilibria; Oxidation products of alcohols	**Chains and Rings and Spectroscopy** Arenes – carbonyl compounds; Carboxylic acids & derivatives Nitrogen compounds Stereoisomerism; Polymerization; Spectroscopy
Unit 5	1 h 30 min　　　　A 15% Structured questions	Internal assessment　A15% + Structured/free response	45 min　　　　A 15% Structured questions + 45 min Structured questions
	Redox equilibria; Transition Metals; Organic III Kinetics III – quantitative Organic IV – analysis, synthesis and applications	**Investigations and Applications** Practical + **Special Studies**: *one of* Biochemistry Chemical Engineering Food Science Materials Science Mineral Process Chemistry	**Trends and Patterns** Lattice Enthalpy Group III; Transition metals + **Options**: *one of* E2 Biochemistry E3 Environmental Chemistry E4 Methods of Analysis & Detection E5 Gases, Liquids and Solids E6 Transition Elements
Unit 6	Internal assessment　A 20% or practical test Data Interpretation & extended writing	2 h　　　　A 20% Structured questions (+ Synoptic)	1 h 30 min　　　　A 20% Structured questions + Coursework or 2 h practical test
	Whole course	**Applying Chemical Principles** Born–Haber cycle; Structure and bonding; Redox equilibria; Natural and synthetic polymers; Transition elements; Organic synthesis; Instrumental methods **Whole course**	**Unifying Concepts** **whole course** + **Experimental Skills**

	OCR (Salters)		AQA	
AS	AS 3887		AS 5421	
Unit 1	1 h 30 min	AS 30% A 15%	1 h 30 min	AS 30% A 15%
	Structured questions		Structured questions + short answer	
	Chemistry for Life Elements of life Developing fuels		**Foundation Chemistry** Atomic structure; Bonding; Periodicity	
Unit 2	2 h	AS 40% A 20%	1 h 30 min	AS 30% A 15%
	Structured questions		Structured questions and short answer	
	Minerals to Medicines Minerals to elements; Atmosphere; Polymer revolution; What's in a Medicine?		**Foundation Chemistry 2** (physical and inorganic) Energetics; Kinetics; Equilibria; Redox Halogens Extraction of metals	
Unit 3	2 weeks	AS 15% A 7.5%	1 h 15 min	AS 25% A 12.5%
	+ Internal assessment	AS 15% A 7.5%	Structured questions + Centre-assessed coursework or 2 h practical test	
	Skills for Chemistry **Open-book paper** **and** **Experimental skills**		**Foundation Chemistry 3** (Introduction to organic) Organic nomenclature; Petroleum and alkanes; Alkenes and epoxyethane Haloalkanes; Alcohols **+ Experimental**	
A2	A7887		A 6421	
Unit 4	1 h 30 min	A 15%	1 h 30 min	A 15%
	Structured questions		Structured questions	
	Polymers, Proteins and Steel Designer polymers; Engineering proteins; Steel story		**Further Chemistry I** (physical and organic) Kinetics; Equilibria; Acids and bases; Isomerism Carbonyl group; Aromatic chem.; Amines; Amino acids; Polymers; Synthesis; Structure determination	
Unit 5	2 h	A 20%	2 h	A 20%
	Structured questions		Structured data analysis, comprehension including synoptic questions	
	Chemistry by Design Aspects of agriculture; Colour by design; The oceans; Medicines by design; Visiting the chemical industry		**Further Chemistry 2** (thermodynamics and inorganic) Thermodynamics; Periodicity; Redox Transition metals; Inorganic aqueous reactions	
Unit 6	Internal assessment	A 15%	1 h	A 10%
			Synoptic – objective questions + Centre-assessed coursework or 2 h practical examination	A 5%
	Individual Investigation		**Further Chemistry 3** **Whole course +** **Experimental work**	

How to revise

There is no one method of revising which works for everyone. It is therefore important to discover the approach that suits you best. The following rules may serve as general guidelines.

GIVE YOURSELF PLENTY OF TIME

Leaving everything until the last minute reduces your chances of success. Work will become more stressful which will reduce your concentration. There are very few people who can revise everything 'the night before' and still do well in an examination the next day.

PLAN YOUR REVISION TIMETABLE

You need to plan your revision timetable some weeks before the examination and make sure that your time is shared suitably between all your subjects. Once you have done this, follow it – don't be side-tracked. Stick your timetable somewhere prominent where you will keep seeing it.

RELAX

Concentrated revision is very hard work. It is as important to give yourself time to relax as it is to work. Build some leisure time into your revision timetable.

GIVE YOURSELF A BREAK

When you are working, work for about an hour then take a short tea or coffee break for 15 to 20 minutes.
Then go back to another productive revision period.

FIND A QUIET CORNER

Find the conditions in which you can revise most efficiently. Many people think they can revise in a noisy busy atmosphere – most cannot! Any distraction lowers concentration. Revising in front of a television doesn't generally work!

KEEP TRACK

Use checklists and the Examination Board specification to keep track of your progress. Mark off topics you have revised and feel confident with. Concentrate your revision on things you are less happy with.

MAKE SHORT NOTES, USE COLOURS

Revision is often more effective when you do something active rather than simply reading material. As you read through your notes and textbooks make brief notes on key ideas. If this book is your own property you could highlight the parts of pages that are relevant to the specification you are following. Concentrate on understanding the ideas rather than just memorising the facts.

PRACTISE ANSWERING QUESTIONS

As you finish each topic, try answering some questions. There are some in this book to help you. You should also use questions from past papers. At first you may need to refer to notes or textbooks. As you gain confidence you will be able to attempt questions unaided, just as you will in the exam.

ADJUST YOUR LIFESTYLE

Make sure that any paid employment and leisure activities allow you adequate time to revise. There is often great temptation to increase the time spent in paid employment when it is available. This can interfere with a revision timetable and make you too tired to revise thoroughly. Consider carefully whether the short-term gains of paid employment are preferable to the long-term rewards of examination success.

Revision Timetable

Week beginning ....

Week no. .... **Exams in** .... **weeks!**

	Monday	Tuesday	Wednesday	Thursday	Friday	Saturday	Sunday
Morning							
Afternoon							
Evening							

Have you allowed time to:
- fulfil any employment commitments?
- complete outstanding coursework?
- meet with friends or undertake other social activities?
- take a break during a revision session?

Timetable check: topics that need extra work

1 ....
2 ....
3 ....
4 ....
5 ....

Success in examinations

Examination technique

The following are some points to note when taking an examination.

- Read the question carefully. Make sure you understand exactly what is required.
- If you find that you are unable to do a part of a question do not give up. The next part may be easier and may provide a clue to what you might have done in the part you found difficult.
- Note the number of marks per question as a guide to the depth of response needed (see below).
- Underline or note the key words that tell you what is required (see opposite).
- Underline or note data as you read the question.
- Structure your answers carefully.
- Show all steps in calculations. Include equations you use and show the substitution of data. Remember to work in SI units.
- Make sure your answers are to suitable significant figures (usually 2 or 3) and include a unit.
- Consider whether the magnitude of a numerical answer is reasonable for the context. If it is not check your working.
- Draw diagrams and graphs carefully.
- Read data from graphs carefully; note scales and prefixes on axes.
- Keep your eye on the clock but do not panic.
- If you have time at the end – use it. Check that your descriptions and explanations make sense. Consider whether there is anything you could add to an explanation or description. Repeat calculations to ensure that you have not made a mistake.

Depth of response

Look at the **marks allocated to the question**.

This is usually a good guide to the depth of the answer required. It also gives you an idea how long to spend on the question. If there are 60 marks available in a 90 minute exam your 1 mark should be earned in 1.5 minutes.

Explanations and descriptions

If a **4 mark** question requires an explanation or description you will need to make **FOUR** distinct relevant points.

You should note, however, that simply mentioning the four points will not necessarily earn full marks. The points need to be made in a coherent way that makes sense and fits the context and demands of the questions.

Calculations

In calculation questions marks will be awarded for method and the final answer.

In a **3 mark** calculation question you **may** obtain all three marks if the final answer is correct even if you show no working. However you should always show your working because

- sometimes the working is required for full marks
- if you make an error in the calculation you cannot gain any method marks unless you have shown your working.

In general in a **3 mark calculation** you earn

1 mark for quoting a relevant equation or using a suitable method

1 mark for correct substitution of data or some progress toward the final answer

1 mark for a correct final answer given to suitable significant figures with a correct unit.

Errors carried forward

If you make a mistake in a calculation and need to use this incorrect answer in a subsequent part of the question you can still gain full marks. Do not give up if you think you have gone wrong. Press on using the data you have.

Key words

How you respond to a question can be helped by studying the following, which are the more common key words used in examination questions.

Key word

Name

Answer is usually a technical term consisting of one or two words.

List

You need to write down a number of points (often a single word) with no elaboration.

Define

The answer is a formal meaning of a particular term.

What is meant by…?

This is often used instead of 'define'.

State

The answer is a word or concise phrase with no elaboration.

Describe

The answer is a description of an effect, experiment, or (e.g.) graph shape. No explanations are required.

Suggest

In your answer you will need to use your knowledge and understanding of topics in the specification to deduce or explain an effect that may be in a novel context. There may be no single correct answer to the question.

Calculate

A numerical answer is to be obtained, usually from data given in the question. Remember to give your answer to a suitable number of significant figures and give a unit.

Determine

Often used instead of 'calculate'. You may need to obtain data from graphs, tables, or measurements.

Explain

The answer will be extended prose. You will need to use your knowledge and understanding of scientific ideas or theories to explain a statement that has been made in the question or earlier in your answer. A question often asks you to 'State and explain'.

Justify

Similar to 'explain'. You will have made a statement and now have to provide a reason for giving that statement.

Draw

Simply draw a diagram. If labelling or a scale drawing is needed you will usually be asked for this but it is sensible to provide labelling even if it is not asked for.

Sketch

This usually relates to a graph. You need to draw the general shape of the graph on labelled axes. You should include enough quantitative detail to show relevant intercepts and/or whether the graph is exponential or some inverse function, for example.

Plot

The answer will be an accurate plot of a graph on graph paper. Often followed by a question asking you to 'determine some quantity from the graph' or to 'explain its shape'.

Estimate

You may need to use your knowledge and/or your experience to deduce the magnitude of some quantities to arrive at the order of magnitude for some other quantity defined in the question.

Discuss

This will require an extended response in which you demonstrate your knowledge and understanding of a given topic.

Show that

You will have been given either a set of data and a final value (that may be approximate) or an algebraic equation. You need to show clearly all basic equations that you use and all the steps that lead to the final answer.

REVISION NOTE

In your revision remember to

- learn the formulae that are not on your formula sheet
- make sure that you know what is represented by all the symbols in equations on your formula sheets.

Carrying out investigations

Keep a notebook

Record
- all your measurements
- any problems you have met
- details of your procedures
- any decisions you have made about apparatus or procedures including those considered and discarded
- relevant things you have read or thoughts you have about the problem.

Define the problem

Write down the aim of your experiment or investigation. Note the variables in the experiment. Define those that you will keep constant and those that will vary.

Suggest a hypothesis

You should be able to suggest the expected outcome of the investigation in the basis of your knowledge and understanding of science. Try to make this as quantitative as you can, justifying your suggestion with equations wherever possible.

Do rough trials

Before commencing the investigation in detail do some rough tests to help you decide on
- suitable aparatus
- suitable procedures
- range and intervals at which you will take measurements.

Consider carefully how you will conduct the experiment in a way that will ensure safety to persons and to equipment. Remember to consider alternative apparatus and procedures and justify your final decision.

Carry out the experiment

Remember all the skills you have learnt during your course:
- note all readings that you make
- take repeats and average whenever possible
- use instruments that provide suitably accurate data
- consider the accuracy of the measurements you are making
- analyse data as you go along so that you can modify the approach or check doubtful data.

Presentation of data

Tabulate all your observations, remembering to
- include the quantity, any prefix, and the unit for the quantity at the head of each column
- include any derived quantities that are suggested by your hypothesis
- quote measurements and derived data to an accuracy/significant figures consistent with your measuring instruments and techniques, and be consistent
- make sure figures are not ambiguous.

Graph drawing

Remember to
- label your axes with quantity and unit
- use a scale that is easy to use and uses the graph paper effectively
- plot points clearly (you may wish to include 'error bars')
- draw the best line through your plotted points
- consider whether the gradient and area under your graph has significance.

Analysing data

This may include
- calculation of a result
- drawing of a graph
- statistical analysis of data
- analysis of uncertainties in the original readings, derived quantities, and results.

Make sure that the stages in the processing of your data are clearly set out.

Evaluation of the investigation

The evaluation should include the following points:
- draw conclusions from the experiment
- identify any systematic errors in the experiment
- comment on your analysis of the uncertainties in the investigation
- review the strengths and weaknesses in the way the experiment was conducted
- suggest alternative approaches that might have improved the experiment in the light of experience.

Use of IT

You may have used data capture techniques when making measurements or used IT in your analysis of data. In your analysis you should consider how well this has performed. You might include answers to the following questions.
- What advantages were gained by the use of IT?
- Did the data capture equipment perform better than you could have achieved by a non-IT approach?
- How well has the data analysis software performed in representing your data graphically, for example?

THE REPORT

Remember that your report will be read by an assessor who will not have watched you doing the experiment. For the most part the assessor will only know what you did by what you write so do not leave out important information.

If you write a good report it should be possible for the reader to repeat what you have done should they wish to check your work.

A **word-processed report** is worth considering. This makes the report much easier to revise if you discover some aspect you have omitted. It is also make it easier for the assessor to read.

NOTE: The report may be used as portfolio evidence for assessment of Application of number, Communication, and IT Key Skills.

Use subheadings

These help break up the report and make it more readable. As a guide the subheadings could be the main sections of the investigation: **aims**, **diagram of apparatus**, **procedure**, etc.

Coping with coursework

TYPES OF COURSEWORK

Coursework takes different forms with different specifications. You may undertake

- short experiments as a routine part of your course
- long practical tasks prescribed by your teacher/lecturer
- a long investigation of a problem decided by you and agreed with your teacher
- a research and analysis exercise using books and IT and other resources.

A short experiment

This may take one or two laboratory sessions to complete and will usually have a specific objective that is closely linked to the topic you are studying at the time.

You may only be assessed on one or two of the skills in any one assessment.

A long investigation

This make take 5 to 10 hours of class time plus associated homework time.

You will probably be assessed on all the skills in a long investigation.

Research and analysis task

This may take a similar amount of time but is likely to be spread over a longer period. This is to give you time to obtain information from a variety of sources. You will be assessed on

- the planning of the research
- the use of a variety of sources of information
- your understanding of what you have discovered
- your ability to identify and evaluate relevant information
- the communication of your findings in writing or in an oral presentation.

Make sure you know in detail what is expected of you in the course you are following. A summary appears on the previous pages.

STUDY THE CRITERIA

Each Examination Board produces criteria for the assessment of coursework. The practical skills assessed are common to all Boards but the way each skill is rewarded is different for each specification. Ensure that you have a copy of the assessment criteria so that you know what you are trying to achieve and how your work will be marked.

PLAN YOUR TIME

Meeting the deadline is often a major problem in coping with coursework.

Do not leave all the writing-up to the end.

Using a word processor you can draft the report as you go along.

You can then go back and tidy it up at the end.

Draw up an initial plan

Include the following elements:

The aim of the project

What are you going to investigate practically?
or
What is the topic of your research?

A list of resources

What are your first thoughts on apparatus?
or
Where are you going to look for information?
(Books; CD ROMs; Internet)
or
Is there some organisation you could write to for information?

Theoretical ideas

What does theory suggest will be the outcome?
or
What are the main theoretical ideas that are linked with your investigation or research project?

Timetable

What is the deadline?
What is your timetable for

Laboratory tasks
How many lab sessions are there?
Initial thoughts on how they are to be used

Non-laboratory tasks
Initial analysis of data
Writing up or word processing part of your final report
Making good diagrams of your apparatus
Revising your time plan
Evaluating your data or procedures

Key Skills

What are Key Skills?

These are skills that are not specific to any subject but are general skills that enable you to operate competently and flexibly in your chosen career. Visit the Key Skills website (www.keyskillssupport.net) or phone the Key Skills help line to obtain full, up-to-date information.

While studying your AS or A level courses you should be able to gather evidence to demonstrate that you have achieved competence in the Key Skills areas of

- **Communication**
- **Application of Number**
- **Information Technology**

You may also be able to prove competence in three other key skills areas:

- **Working with Others**
- **Improving Your Own Learning**
- **Problem Solving**

Only the first three will be considered here and only an outline of what you must do is included. You should obtain details of what you need to know and be able to do. You should be able to obtain this from your examination centre.

Communication

You must be able to

- create opportunities for others to contribute to group discussions about complex subjects
- make a presentation using a range of techniques to engage the audience
- read and synthesise information from extended documents about a complex subject
- organise information coherently, selecting a form and style of writing appropriate to complex subject matter.

Application of Number

You must be able to plan and carry through a substantial and complex activity that requires you to

- plan your approach to obtaining and using information, choose appropriate methods for obtaining the results you need, and justify your choice
- carry out multistage calculations including use of a large data set (over 50 items) and re-arrangement of formulae
- justify the choice of presentation methods and explain the results of your calculations.

Information Technology

You must be able to plan and carry through a substantial activity that requires you to

- plan and use different sources and appropriate techniques to search for and select information based on judgement of relevance and quality
- use automated routines to enter and bring together information, and create and use appropriate methods to explore, develop, and exchange information
- develop the structure and content of your presentation, using others' views to guide refinements, and information from different sources.

A **complex subject** is one in which there are a number of ideas, some of which may be abstract and very detailed. Lines of reasoning may not be immediately clear. There is a requirement to come to terms with specialised vocabulary. A **substantial activity** is one that includes a number of related tasks. The result of one task will affect the carrying out of others. You will need to obtain and interpret information and use this to perform calculations and draw conclusions.

What standard should you aim for?

Key skills are awarded at four levels (1–4). In your A level courses you will have opportunities to show that you have reached **level 3** but you could produce evidence that demonstrates that you are competent at a higher level. You may achieve a different level in each Key Skill area.

What do you have to do?

You need to show that you have the necessary underpinning knowledge in the Key Skills area and produce evidence that you are able to apply this in your day-to-day work. You do this by producing a portfolio that contains

- evidence in the form of reports when it is possible to provide written evidence
- evidence in the form of assessments made by your teacher when evidence is gained by observation of your performance in the classroom or laboratory

The evidence may come from only one subject that you are studying but it is more likely that you will use evidence from all of your subjects.

It is up to you to produce the best evidence that you can.

The specifications you are working with in your AS or A level studies will include some ideas about the activities that form part of your course and can be used to provide this evidence. Some general ideas are summarised below but refer to the specification for more detail.

Communication

In Science you could achieve this by

- undertaking a long practical or research investigation on a complex topic (e.g. Use of nuclear radiation in medicine)
- writing a report based on your experimentation or research using a variety of sources (books, magazines, CD ROMs, Internet, newspapers)
- making a presentation to your fellow students
- using a presentation style that promotes discussion or criticism of your findings, enabling others to contribute to a discussion that you lead.

Application of Number

In Science you could achieve this by

- undertaking a long investigation or research project that requires details planning of methodology
- considering alternative approaches to the work and justifying the chosen approach
- gathering sufficient data to enable analysis by statistical and graphical methods
- explaining why you analysed the data as you did
- drawing the conclusions reached as a result of your investigation.

Information Technology

In Science you could achieve this by

- using CD ROMs and the Internet to research a topic
- identifying those sources which are relevant
- identifying where there is contradictory information and identifying which is most probably correct
- using a word processor to present your report, drawing in relevant quotations from the information you have gathered
- using a spreadsheet to analyse data that you have collected
- using data capture techniques to gather information and mathematics software to analyse the data.

Answering the question

Most of the Unit tests for AS and A2 consist of Structured Questions.

Be guided by the mark allocation – in general, 2 marks will require two distinct points.

Remember that you are given plenty of time to read the question carefully.

Be prepared to state the obvious (there may well be marks for it).

Structured questions are designed to test a range of syllabus topics within one overall theme.

Q (a) Define the term 'standard enthalpy of combustion'. (3 marks)

(b) Draw a reaction profile diagram for the combustion of methane (CH_4). (3 marks)

(c) Given $\Delta H^{\ominus}_{comb}$ (CH_4) = –890 kJ mol^{-1} calculate the energy released on burning 1.00 dm^3 methane. (Under the conditions of the experiment 1 mole of gas occupies 24 dm^3.) (2 marks)

A (a) The enthalpy change when 1 mole of the substance is completely burned (in air/oxygen) under standard conditions, 1 atmosphere pressure, 298 K temperature.
- 1 mole ✔
- completely burned ✔
- standard conditions given ✔

(b)

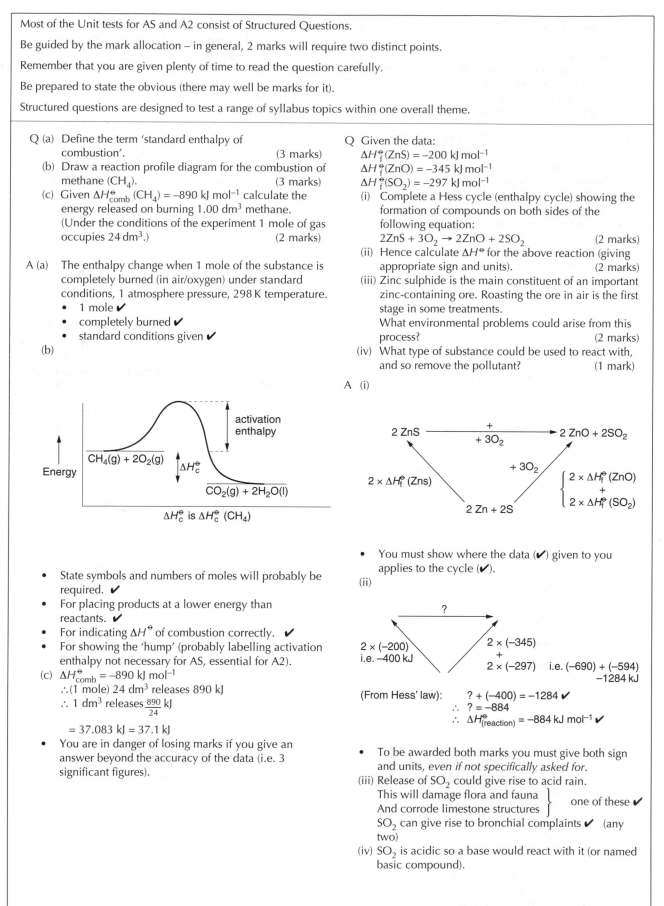

$\Delta H^{\ominus}_c$ is $\Delta H^{\ominus}_c$ (CH_4)

- State symbols and numbers of moles will probably be required. ✔
- For placing products at a lower energy than reactants. ✔
- For indicating $\Delta H^{\ominus}$ of combustion correctly. ✔
- For showing the 'hump' (probably labelling activation enthalpy not necessary for AS, essential for A2).

(c) $\Delta H^{\ominus}_{comb}$ = –890 kJ mol^{-1}
∴ (1 mole) 24 dm^3 releases 890 kJ
∴ 1 dm^3 releases $\frac{890}{24}$ kJ

= 37.083 kJ = 37.1 kJ

- You are in danger of losing marks if you give an answer beyond the accuracy of the data (i.e. 3 significant figures).

Q Given the data:
$\Delta H^{\ominus}_f$(ZnS) = –200 kJ mol^{-1}
$\Delta H^{\ominus}_f$(ZnO) = –345 kJ mol^{-1}
$\Delta H^{\ominus}_f$(SO$_2$) = –297 kJ mol^{-1}

(i) Complete a Hess cycle (enthalpy cycle) showing the formation of compounds on both sides of the following equation:
$2ZnS + 3O_2 \rightarrow 2ZnO + 2SO_2$ (2 marks)

(ii) Hence calculate $\Delta H^{\ominus}$ for the above reaction (giving appropriate sign and units). (2 marks)

(iii) Zinc sulphide is the main constituent of an important zinc-containing ore. Roasting the ore in air is the first stage in some treatments.
What environmental problems could arise from this process? (2 marks)

(iv) What type of substance could be used to react with, and so remove the pollutant? (1 mark)

A (i)

- You must show where the data (✔) given to you applies to the cycle (✔).

(ii)

(From Hess' law): ? + (–400) = –1284 ✔
∴ ? = –884
∴ $\Delta H^{\ominus}_{(reaction)}$ = –884 kJ mol^{-1} ✔

- To be awarded both marks you must give both sign and units, *even if not specifically asked for*.

(iii) Release of SO$_2$ could give rise to acid rain.
This will damage flora and fauna } one of these ✔
And corrode limestone structures }
SO$_2$ can give rise to bronchial complaints ✔ (any two)

(iv) SO$_2$ is acidic so a base would react with it (or named basic compound).

Answering the question

Q (i) Complete the balanced equation for the formation of poly(chloroethene) from chlorethene:
$$n \; C_2H_3Cl \rightarrow$$
(1 mark)

(ii) How does intermolecular bonding account for the rigidity of poly(chloroethene) (PVC)? (2 marks)

(iii) Explain how the use of plasticisers modifies the properties of poly(chloroethene). (2 marks)

A possible route for making chloroethene in the laboratory is given below:

	Stage 1		Stage 2	
C_2H_4	$\rightarrow$	$C_2H_4Cl_2$	$\rightarrow$	C_2H_3Cl
ethene		1,2-dichloroethene		chloroethene

(iv) What is the name given to the type of reaction in stage 1? (1 mark)

(v) Describe what is happening in stage 2 and suggest a reagent to bring this change about. (2 marks)

(vi) 1,2-dichloroethane can be used to make ethane-1,2-diol for the polyester industry. Suggest a reagent and conditions for this reaction. (2 marks)

(vii) There are three isomers of dichloroethene $C_2H_2Cl_2$. Give structural (graphical/displayed) formulae for them, and names. (3 marks)

(viii) The compound 1-bromo-1-chloroethane ($CH_3CHBrCl$) has a chiral centre. Draw diagrams to show the two optical isomers. (2 marks)

A (i) $n \; C_2H_3Cl \rightarrow (C_2H_3Cl)_n$
But a better answer would be:

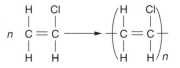

(ii) The C—Cl is permanently polar:
There are permanent dipole to permanent dipole (✔) intermolecular bonds between polymer chains. This prevents them readily sliding over each other, making the material rigid (✔).

• Note that in fully answering this question, rigidity must be referred to.

(iii) Plasticiser molecules keep chains further apart, thus diminishing intermolecular bonding (✔). The material is more flexible (or less rigid) (✔).

• Note that attention must be paid both to effect and cause.

(iv) (Electrophilic) addition (✔).

(v) There is a loss of HCl (✔) (an elimination). A base such as NaOH(aq) might bring this about (in practice usually $Ca(OH)_2$).

• Inspection of product and reactant gives the first mark.

(vi) • If in any doubt always write out the structural formulae of compounds involved.
i.e. $ClCH_2CH_2Cl \rightarrow HOCH_2CH_2OH$

This is obviously a (nucleophilic) substitution by an OH^- ion.

The ideal reagent would be aqueous NaOH (✔) and the reaction will proceed at room conditions (✔).

(vii)

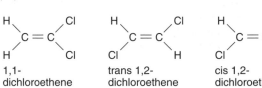

• In questions such as this give full structural diagrams. The presence of the double bond should make you consider geometric isomerism.

(viii)

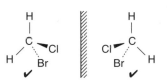

• The three-dimensional system is essential to score marks. The isomers must be mirror images.

Answering the question – extended writing

All examinations will include some questions that require longer written answers. Some marks will be specifically awarded for **quality of the written communication**.

Q Explain, in terms of polymer chains, why nylon-6,6 is a strong polymeric material at room temperature, but becomes less strong as the temperature is raised.

A The structure of nylon-6.6 is that of long chains containing the repeating unit:

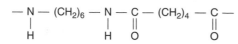

Both the N—H bond and the C=O bond are permanently polar:

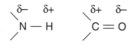

This means that in addition to weak Van der Waals bonds between the hydrocarbon parts of the molecule there can be stronger hydrogen bonding at regular intervals between molecules.

This makes it difficult for chains to move over each other, therefore the plastic is strong. At higher temperatures there is sufficient energy to overcome the hydrogen bonds, making movement of chains easier, and reducing the strength of the material.

• Note the use of formulae and diagrams although not actually asked for. These aid clarity of communication. The answer is quite detailed in order to make the meaning clear.

Q Describe and explain the reactivity of chlorine, bromine, and iodine as shown in their displacement reactions with aqueous solutions of halides. You should include relevant equations.

A The reactivity of the halogens decreases as you go down the group. Of the three, chlorine, the one with the smallest atoms, has the greatest attraction for electrons. This is because the quantum shell that the outer electrons are in is close to the nucleus. It therefore most readily forms a halide ion, accepting an electron. Chlorine therefore displaces both bromine and iodine from solutions of their ions.

$Cl_2(aq) + 2Br^-(aq) \rightarrow 2Cl^-(aq) + Br_2(aq)$

$Cl_2(aq) + 2I^-(aq) \rightarrow 2Cl^-(aq) + I_2(aq)$

Using the same argument: bromine forms a halide ion more readily than iodine and so displaces iodine from a solution of its ions.

$Br_2(aq) + 2I^-(aq) \rightarrow 2Br^-(aq) + I_2(aq)$

• NB More than one equation is needed here. For a complete answer it is necessary to consider both chlorine and bromine. It is important to start the answer with the overall picture.

Answering the question – laboratory situations

A feature of chemistry examinations – all Boards – is the requirement to illustrate an answer with a diagram of suitable apparatus.

Generally the question asks for 'a fully labelled diagram'. It is prudent to supply labels in any event.

The most frequently encountered types of question are given below as examples.

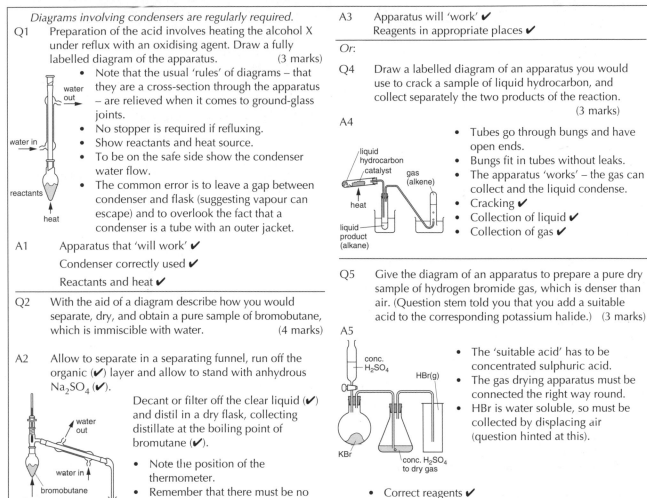

Diagrams involving condensers are regularly required.

Q1 Preparation of the acid involves heating the alcohol X under reflux with an oxidising agent. Draw a fully labelled diagram of the apparatus. (3 marks)

- Note that the usual 'rules' of diagrams – that they are a cross-section through the apparatus – are relieved when it comes to ground-glass joints.
- No stopper is required if refluxing.
- Show reactants and heat source.
- To be on the safe side show the condenser water flow.
- The common error is to leave a gap between condenser and flask (suggesting vapour can escape) and to overlook the fact that a condenser is a tube with an outer jacket.

A1 Apparatus that 'will work' ✔

Condenser correctly used ✔

Reactants and heat ✔

Q2 With the aid of a diagram describe how you would separate, dry, and obtain a pure sample of bromobutane, which is immiscible with water. (4 marks)

A2 Allow to separate in a separating funnel, run off the organic (✔) layer and allow to stand with anhydrous Na_2SO_4 (✔).

Decant or filter off the clear liquid (✔) and distil in a dry flask, collecting distillate at the boiling point of bromutane (✔).

- Note the position of the thermometer.
- Remember that there must be no gaps at joints.
- Essential labels are shown.

Some Boards may require you to design your own apparatus. Frequently this involves manipulating gases. What is expected will be simple – but it must work and use apparatus you are familiar with.
i.e.

Q3 A solution of hydrogen peroxide (H_2O_2) readily decomposes in contact with manganese dioxide powder to give oxygen gas. Draw a labelled diagram of an apparatus for collection of a few test tubes of oxygen gas. (2 marks)

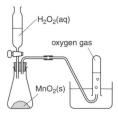

- Essentially this is a matter of recognising that apparatus must be assembled before any gas is evolved – hence the use of a tap funnel.
- Tubing must be seen to go through bungs (and not have closed ends).
- Bungs must be 'air-tight'.
- All reagents and essential products must be labelled.

A3 Apparatus will 'work' ✔
Reagents in appropriate places ✔

Or:

Q4 Draw a labelled diagram of an apparatus you would use to crack a sample of liquid hydrocarbon, and collect separately the two products of the reaction. (3 marks)

A4

- Tubes go through bungs and have open ends.
- Bungs fit in tubes without leaks.
- The apparatus 'works' – the gas can collect and the liquid condense.
- Cracking ✔
- Collection of liquid ✔
- Collection of gas ✔

Q5 Give the diagram of an apparatus to prepare a pure dry sample of hydrogen bromide gas, which is denser than air. (Question stem told you that you add a suitable acid to the corresponding potassium halide.) (3 marks)

A5

- The 'suitable acid' has to be concentrated sulphuric acid.
- The gas drying apparatus must be connected the right way round.
- HBr is water soluble, so must be collected by displacing air (question hinted at this).

- Correct reagents ✔
- Apparatus that works (no leaks, etc.) ✔
- Correct drying agent or correct collection method ✔

Note that written questions will test your knowledge and practical experience in those areas unlikely to be applicable to practical examinations or coursework.

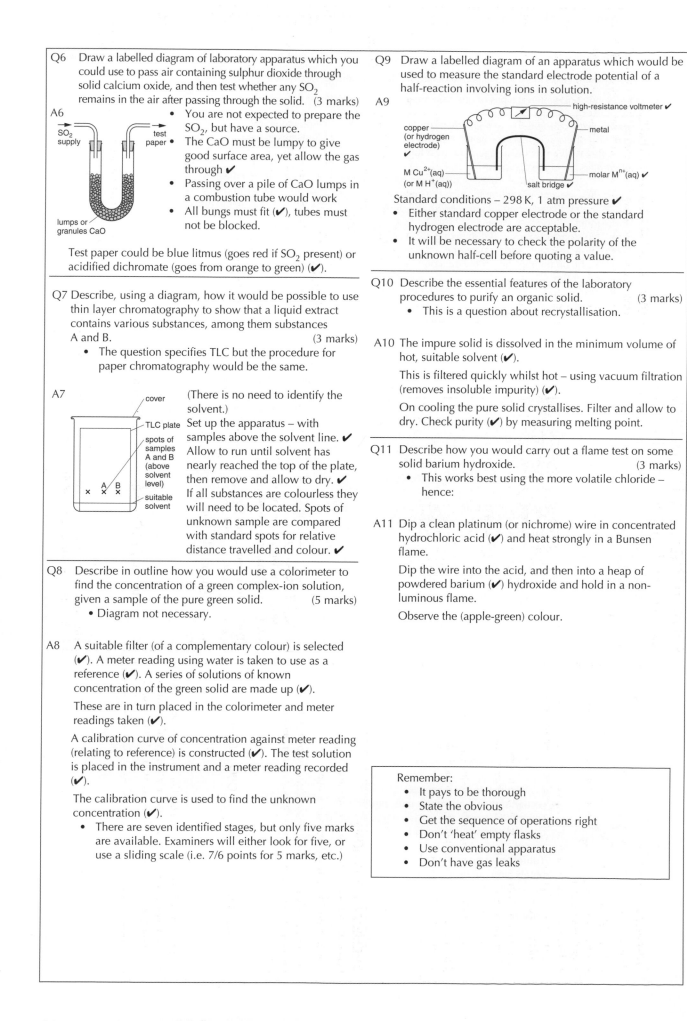

Q6 Draw a labelled diagram of laboratory apparatus which you could use to pass air containing sulphur dioxide through solid calcium oxide, and then test whether any SO_2 remains in the air after passing through the solid. (3 marks)

A6

SO₂ supply

test paper

lumps or granules CaO

- You are not expected to prepare the SO_2, but have a source.
- The CaO must be lumpy to give good surface area, yet allow the gas through ✔
- Passing over a pile of CaO lumps in a combustion tube would work
- All bungs must fit (✔), tubes must not be blocked.

Test paper could be blue litmus (goes red if SO_2 present) or acidified dichromate (goes from orange to green) (✔).

Q7 Describe, using a diagram, how it would be possible to use thin layer chromatography to show that a liquid extract contains various substances, among them substances A and B. (3 marks)
- The question specifies TLC but the procedure for paper chromatography would be the same.

A7

cover
TLC plate
spots of samples A and B (above solvent level)
suitable solvent
A B

(There is no need to identify the solvent.) Set up the apparatus – with samples above the solvent line. ✔ Allow to run until solvent has nearly reached the top of the plate, then remove and allow to dry. ✔ If all substances are colourless they will need to be located. Spots of unknown sample are compared with standard spots for relative distance travelled and colour. ✔

Q8 Describe in outline how you would use a colorimeter to find the concentration of a green complex-ion solution, given a sample of the pure green solid. (5 marks)
- Diagram not necessary.

A8 A suitable filter (of a complementary colour) is selected (✔). A meter reading using water is taken to use as a reference (✔). A series of solutions of known concentration of the green solid are made up (✔).

These are in turn placed in the colorimeter and meter readings taken (✔).

A calibration curve of concentration against meter reading (relating to reference) is constructed (✔). The test solution is placed in the instrument and a meter reading recorded (✔).

The calibration curve is used to find the unknown concentration (✔).
- There are seven identified stages, but only five marks are available. Examiners will either look for five, or use a sliding scale (i.e. 7/6 points for 5 marks, etc.)

Q9 Draw a labelled diagram of an apparatus which would be used to measure the standard electrode potential of a half-reaction involving ions in solution.

A9

high-resistance voltmeter ✔
copper (or hydrogen electrode) ✔
metal
M Cu²⁺(aq) (or M H⁺(aq))
salt bridge ✔
molar Mⁿ⁺(aq) ✔

Standard conditions – 298 K, 1 atm pressure ✔
- Either standard copper electrode or the standard hydrogen electrode are acceptable.
- It will be necessary to check the polarity of the unknown half-cell before quoting a value.

Q10 Describe the essential features of the laboratory procedures to purify an organic solid. (3 marks)
- This is a question about recrystallisation.

A10 The impure solid is dissolved in the minimum volume of hot, suitable solvent (✔).

This is filtered quickly whilst hot – using vacuum filtration (removes insoluble impurity) (✔).

On cooling the pure solid crystallises. Filter and allow to dry. Check purity (✔) by measuring melting point.

Q11 Describe how you would carry out a flame test on some solid barium hydroxide. (3 marks)
- This works best using the more volatile chloride – hence:

A11 Dip a clean platinum (or nichrome) wire in concentrated hydrochloric acid (✔) and heat strongly in a Bunsen flame.

Dip the wire into the acid, and then into a heap of powdered barium (✔) hydroxide and hold in a non-luminous flame.

Observe the (apple-green) colour.

Remember:
- It pays to be thorough
- State the obvious
- Get the sequence of operations right
- Don't 'heat' empty flasks
- Use conventional apparatus
- Don't have gas leaks

Quantitative chemistry is

about chemical equations and what they tell you in terms of the amounts of reactants used up and products made

DIFFERENT KINDS OF FORMULA

Empirical formula shows the simplest whole number ratio of atoms in the particles of the substance, e.g. C_2H_6O and CH_2O.

Molecular formula shows the actual number of atoms in a particle of the substance, e.g. C_2H_6O and $C_6H_{12}O_6$.

Structural formula shows the arrangement of atoms in the particle

either written as e.g. CH_3CH_2OH (sometimes called the short form) or drawn to show all bonds as

$$H - \overset{\overset{\displaystyle H}{|}}{\underset{\underset{\displaystyle H}{|}}{C}} - \overset{\overset{\displaystyle H}{|}}{\underset{\underset{\displaystyle H}{|}}{C}} - O - H$$

Some Boards may refer to this as a **displayed** formula, others as a **graphical** formula.

A **skeletal** formula omits the C and H atoms:

(ethanol)

COOH

OH

FULL EQUATIONS

These are used when the stoichiometry of the reaction is being studied. They are concerned with the relative amounts of the reactants used and products made.

State symbols are usually not essential here, although greater credit is given for their use.

e.g.

$$2Mg + O_2 \longrightarrow 2MgO$$

This equation tells us that 2 moles of magnesium react with 1 mole of oxygen molecules to make 1 mole of magnesium oxide.

$$HCl(aq) + Na\,OH(aq)$$
$$\longrightarrow H_2O(l) + NaCl(aq)$$

IONIC EQUATIONS

These are used when we think about how one lot of substances is changed into another.

They are concerned with the bonding, structure, shape, or size of the particles and the mechanism of the reaction.

When writing particle equations state symbols are used, e.g.

$$Cu^{2+}(aq) + 2OH^-(aq) \longrightarrow Cu(OH)_2(s)$$

This equation tells us that a copper aquo ion reacts with two hydroxide ions to make an insoluble product.

CHEMICAL CHANGE

Chemical changes have three main features:

- *New substances* are made
- There is an *energy change* between the reacting system and its surroundings
- There is a fixed relationship between the masses of the reactants and products — this is called the **stoichiometry** of the reaction

Stoichiometry is the name given to the property of pure substances to react together in *whole number ratios of particles*.

Chemical changes are nearly always written as equations showing the reactants and products symbolically in the form of some kind of **formula**.

CHEMICAL EQUATIONS

Reactants are normally written on the left.

Products are normally written on the right.

The arrow → between them means *reacts to give* and sometimes has the conditions written above or below it. e.g.

$$CuCO_3 \xrightarrow{\text{heat}} CuO + CO_2$$

There are two different kinds of equation and although they are often used interchangeably, they really have different uses depending on which feature of the reaction is being studied:

Calculations from equations

KEY RELATIONSHIPS

In the laboratory, substances are most conveniently measured out by weighing for solids and by volume for liquids and gases.

The relationships between amount of substance, number of particles, mass of solid, and volume of gas are very important:

amount   number of particles   mass of solid   volume of gas

$1 \text{ mole} = 6.02 \times 10^{23} = A_r \text{ or } M_r \text{ in grams} = {}^*22.4 \text{ dm}^3 \text{ at s.t.p.}$

Many calculations involve converting from one part of this relationship to another; always go back to this key line at the start of your calculation.

* Standard temperature and pressure are 273 K and 1 atmosphere (101 325 Pa). Often room temperature, 298 K is used: at room temperature a mole of any gas has a volume of 24 dm³.

In electrolysis, the amount of charge involved in the reaction at the electrodes is important:

1 mole of electrons = 96 500 coulombs = 1 Faraday

Another key relationship is given by the ideal gas equation:

$pV = nRT$

where p = pressure in Pa
V = volume in m³
n = number of moles
R = universal gas constant J mol⁻¹ K⁻¹
T = Temperature in K

CALCULATIONS FROM CHEMICAL EQUATIONS

Always try to work through the following steps in this order:

1. write down the equation for the reaction;
2. work out the number of moles of the substance whose amount/mass/volume is given;
3. from the equation, read off the mole ratios (the stoichiometry);
4. using this ratio, work out the number of moles of the unknown substance;
5. using the key relationships above, convert the moles into the units asked for;
6. give your answer to 3 significant figures and remember to put in the units.

$HCl(aq) + Na\,OH(aq) \longrightarrow H_2O(l) + NaCl(aq)$

$[H^+(aq)] = 1.0 mol\, dm^{-3}$
mass of NaOH = 80g
moles of NaOH = ?

The mole is the unit in which *amounts of substance* are measured in chemistry.

The mole is defined as *that amount of substance that contains the same number of particles as there are atoms in exactly 12 g of the isotope carbon-12.*

The number of particles in a mole is found to be 6.02×10^{23}: this number is called the Avogadro constant and has the symbol *L*.

WHEN DOING CALCULATIONS REMEMBER

1. To define the particles you are talking about

Is your mole of oxygen 6.02×10^{23} oxygen atoms which weigh 16 g or 6.02×10^{23} oxygen molecules which weigh 32 g?

2. Substances are often not pure, but are diluted in solutions

The quantity of substance in a solution is called its **concentration**.

Concentration can be expressed in several different ways:

grams per litre shortened to g/l or g l⁻¹
grams per cubic decimetre shortened to g/dm³
moles per litre shortened to mol/l or mol l⁻¹
moles per cubic decimetre shortened to mol/dm³ or mol dm⁻³
molar shortened to M where 1 M means 1 mol dm⁻³

3. Volumes are measured in several different units

1 cubic decimetre ≡ 1 litre ≡ 1000 cubic centimetres

4. Pressure is measured in several different units

1 atmosphere = 101 325 pascal (Pa) (simplified to 10⁵ Pa)
1 Pa = 1 N m⁻²
also 1 atmosphere = 760 mmHg

Spectroscopy gives
us ways of investigating the structure of substances by looking at their spectra

MASS SPECTROMETRY

Description

Particles are bombarded with electrons, which knock other electrons out of the particles making positive ions. The ions are accelerated in an electric field forming an ion beam. The particles in this beam can be sorted according to their masses using an electric field.

Uses

- to measure relative atomic masses
- to find the relative abundance of isotopes in a sample of an element
- to examine the fragments that a molecule might break into so that the identity of the molecule can be found

ULTRA-VIOLET, VISIBLE, INFRA-RED, AND NUCLEAR MAGNETIC RESONANCE SPECTROSCOPY

Description

Energy, in the form of electromagnetic radiation, is applied to the sample. Either the energy taken in by the sample or the energy it gives out is studied.

SAMPLE $\xrightarrow{\text{apply energy}}$ EXCITED SAMPLE $\xrightarrow{\text{energy given out}}$ SAMPLE

see what wavelengths are absorbed
ABSORPTION SPECTRA

see what wavelengths are emitted
EMISSION SPECTRA

The energy of different parts of the electromagnetic spectrum is related to the frequency of that part of the spectrum by the equation

$$E = h\nu,$$ where E is the energy, h is a constant (Planck's constant), and ν is the frequency.

The frequency is related to the wavelength of the radiation by

$$\nu = c/\lambda,$$ where c is the speed of light and λ is the wavelength.

So in summary, the shorter the wavelength, the higher the frequency and the higher the energy.

Different parts of the molecule interact with different wavelengths of radiation. The table below shows how different wavelengths of radiation cause different changes in the particles.

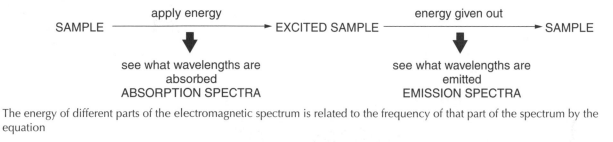

Frequency/MHz	3		3×10^2		3×10^4		3×10^6		3×10^8		3×10^{10}	
Wavelength/m	10^2	10	1	10^{-1}	10^{-2}	10^{-3}	10^{-4}	10^{-5}	10^{-6}	10^{-7}	10^{-8}	10^{-9}
Name of radiation	radio waves			microwaves			infra-red		visible	ultra-violet		X-rays
What happens in the particles	nuclei rotate or spin			molecules rotate			molecules vibrate		electrons in atoms and molecules change orbitals			

USES

Ultra-violet and visible

- to work out electronic structures of atoms and molecules
- indicators in acid/base chemistry
- quantitative analysis in both inorganic and organic chemistry

Infra-red

- detecting the presence of functional groups in organic compounds

Nuclear magnetic resonance

- detecting the number and position of atoms with odd mass numbers in molecules (usually ^{1}H, but also ^{13}C, ^{15}N, ^{19}F, and ^{31}P)

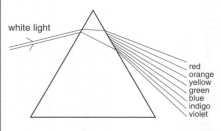

white light

red
orange
yellow
green
blue
indigo
violet

Mass spectra

THE SPECTRUM

The spectrum for elements shows a different peak for each isotope. The height of the peak indicates the amount of each isotope.

The spectrum for compounds will show peaks representing the whole compound particle (this will have the largest mass) and fragments of it which broke up when they were ionised in the electron beam.

Recognition of such fragments can be used in the analysis of organic compounds.

Magnesium spectrum

Carbon dioxide spectrum

THE EXPERIMENTAL SET-UP

A gaseous sample is hit by an electron beam which knocks electrons off the particles making them into positive ions:

$$M(g) + e^- \rightarrow M^+(g) + 2e^-$$

These ions are accelerated in an electric field and aligned into an ion beam. The beam is passed through either an electrostatic field or a magnetic field or both where it is deflected. The deflected particles are then detected and recorded.

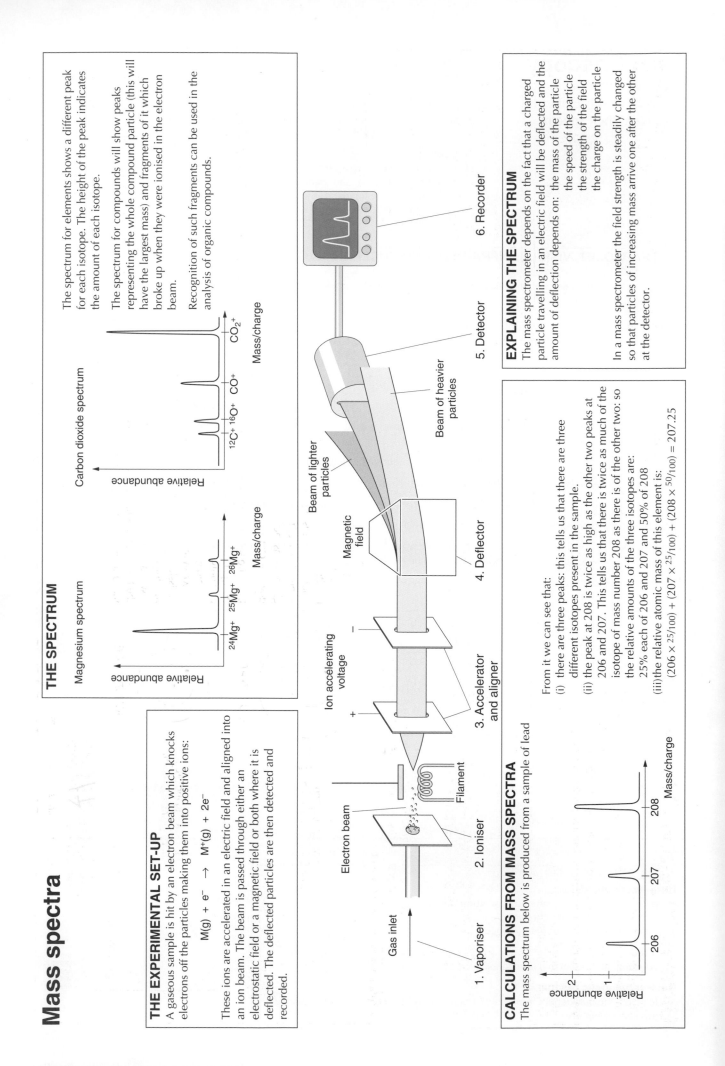

EXPLAINING THE SPECTRUM

The mass spectrometer depends on the fact that a charged particle travelling in an electric field will be deflected and the amount of deflection depends on: the mass of the particle
the speed of the particle
the strength of the field
the charge on the particle

In a mass spectrometer the field strength is steadily changed so that particles of increasing mass arrive one after the other at the detector.

CALCULATIONS FROM MASS SPECTRA

The mass spectrum below is produced from a sample of lead

From it we can see that:

(i) there are three peaks: this tells us that there are three different isotopes present in the sample.

(ii) the peak at 208 is twice as high as the other two peaks at 206 and 207. This tells us that there is twice as much of the isotope of mass number 208 as there is of the other two: so the relative amounts of the three isotopes are:
25% each of 206 and 207 and 50% of 208

(iii) the relative atomic mass of this element is:
$(206 \times {}^{25}/100) + (207 \times {}^{25}/100) + (208 \times {}^{50}/100) = 207.25$

Ultra-violet and visible spectra

The experimental set-up

Emission spectra. A gaseous sample is excited with electrical or thermal energy. Ultra-violet or visible radiation is given out; this is focused into a beam and then split by a prism or diffraction grating; the radiation is then viewed through the telescope or detected photographically.

Absorption spectra. White light from a lamp is directed through a gaseous sample of the substance.

The spectrum

The spectrum produced differs from the normal spectrum of white light in two ways:

(i) it is made up of separate lines (it is discontinuous).

(ii) the lines are in a converging pattern, getting closer as the frequency or energy of the lines increases.

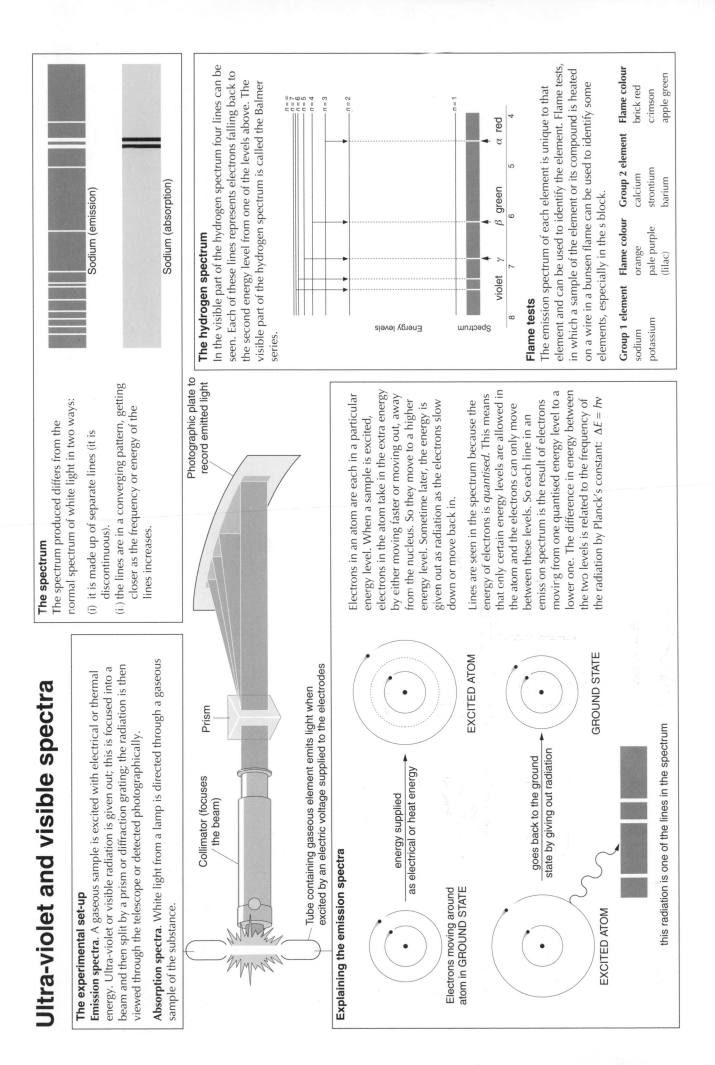

Sodium (emission)

Sodium (absorption)

Photographic plate to record emitted light

Prism

Collimator (focuses the beam)

Tube containing gaseous element emits light when excited by an electric voltage supplied to the electrodes

The hydrogen spectrum

In the visible part of the hydrogen spectrum four lines can be seen. Each of these lines represents electrons falling back to the second energy level from one of the levels above. The visible part of the hydrogen spectrum is called the Balmer series.

$n = \infty$
$n = 7$
$n = 6$
$n = 5$
$n = 4$

$n = 3$

$n = 2$

$n = 1$

Energy levels

Spectrum

α red   4
   5
β green   6
   7
γ
violet   8

Flame tests

The emission spectrum of each element is unique to that element and can be used to identify the element. Flame tests, in which a sample of the element or its compound is heated on a wire in a bunsen flame can be used to identify some elements, especially in the s block.

Group 1 element	Flame colour	Group 2 element	Flame colour
sodium	orange	calcium	brick red
potassium	pale purple (lilac)	strontium	crimson
		barium	apple green

Explaining the emission spectra

Electrons moving around atom in GROUND STATE

energy supplied as electrical or heat energy

EXCITED ATOM

goes back to the ground state by giving out radiation

GROUND STATE

EXCITED ATOM

this radiation is one of the lines in the spectrum

Electrons in an atom are each in a particular energy level. When a sample is excited, electrons in the atom take in the extra energy by either moving faster or moving out, away from the nucleus. So they move to a higher energy level. Sometime later, the energy is given out as radiation as the electrons slow down or move back in.

Lines are seen in the spectrum because the energy of electrons is *quantised*. This means that only certain energy levels are allowed in the atom and the electrons can only move between these levels. So each line in an emission spectrum is the result of electrons moving from one quantised energy level to a lower one. The difference in energy between the two levels is related to the frequency of the radiation by Planck's constant: $\Delta E = h\nu$

EXCITED ATOM

GROUND STATE

Infra-red spectra

Explaining the spectrum

In this kind of spectrometry, infra-red radiation is absorbed, causing the atoms at each end of a bond to vibrate relative to each other. Like a stretched spring between two masses, the energy absorbed by a bond depends on the masses of the atoms and the bond strength. So, as in visible and UV spectra, the vibrational energies are quantised, each kind of bond absorbing its own band of radiation. Only those molecules with charge separation along their bonds absorb in the *infra-red* region and only if this results in a change of dipole moment.

The wavelengths of the energy absorbed, often expressed in **wavenumbers**, appear as dips in the spectrum. Some of these dips indicate the presence of particular functional groups and others are characteristic of the whole molecule.

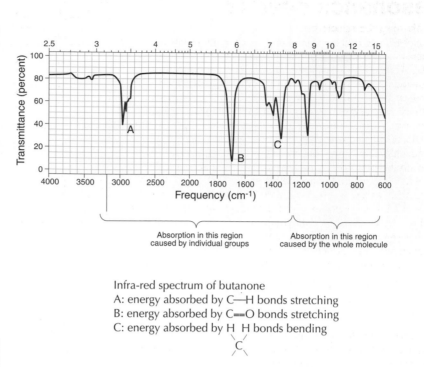

Infra-red spectrum of butanone
A: energy absorbed by C—H bonds stretching
B: energy absorbed by C==O bonds stretching
C: energy absorbed by H H bonds bending
              \ /
               C
              / \

The same group can absorb more than one band of radiation and this makes the final spectrum more complex.

The absorption may be strong (with a big 'dip' in the spectrum) or weak (with a correspondingly smaller dip). The frequency band absorbed is usually quite narrow, but in a few cases (usually involving –OH) is broad.

NB infra-red absorption is characteristic of covalent bonds, including those based on silicon. Consequently glassware and earthenware is used in cookery. Conversely ionic compounds do not absorb and are said to be transparent to infra-red.

ROUTINE ANALYSIS USING IR SPECTROSCOPY

A pure compound absorbs in the infra-red depending upon the various types of covalent bond within it. Its overall spectrum will be distinctive and unique, in other words a 'fingerprint'. Production of the spectrum is relatively quick and analysis can be performed using stored data on computer. For this reason this technique is employed during quality control of organic product, forensic science, air quality monitoring, and as laboratory follow up to the breathalyser test. Peak intensity can yield quantitative information.

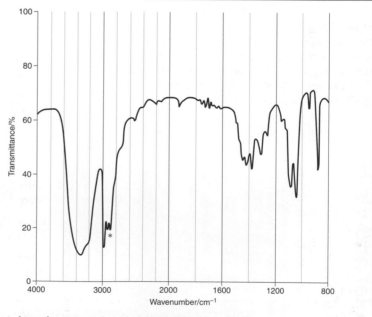

Infra-red spectrum for ethanol. The peak marked * is selected for use in breath analysis.

INFRA-RED SPECTROSCOPY IS A POWERFUL ANALYTICAL TOOL

See also 'Environmental chemistry'

Nuclear magnetic resonance spectra

Explaining the spectrum

Nuclei with an odd number of protons or neutrons have a magnetic moment as though they were spinning in a magnetic field. Normally, there is an equal number of atoms in a sample with each direction of spin and so they cancel each other out. When a strong magnetic field is applied, half the spins align with the field and half against it. This splits the nuclei in terms of energy. Nuclear magnetic resonance, NMR, happens when the nuclei aligned with the field absorb energy and change the direction of their spin. The amount of energy absorbed while they do this depends on the nucleus and its molecular environment and on the magnetic field strength. So NMR can reveal the presence of hydrogen atoms (and other nuclei) in different functional groups. For example, in propanol, there are CH_3—, —CH_2—, and —OH groups and the hydrogens in each of these groups will come into resonance at different frequencies. The frequencies are always measured relative to those for the protons in tetramethylsilane, TMS.

This displacement from the TMS reference is referred to as the chemical shift, and these are characteristic of particular proton environments.

Interpretation of the spectrum is aided by the superimposing of an integrated trace. The relative size of the peaks is computed and plotted. This clearly demonstrates the relative number of protons in the different chemical environments.

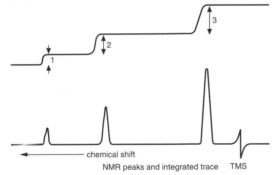

chemical shift

NMR peaks and integrated trace   TMS

NMR IN MEDICINE

Because protons in water, carbohydrates, proteins, and fats represent different environments they give different NMR signals. This allows for different organs in the body to be distinguished. The instrument is called a body scanner, and it produces magnetic resonance images (MRI).

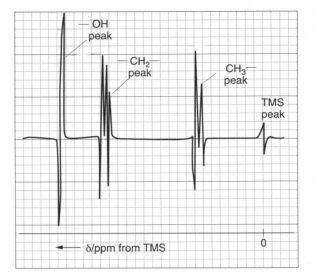

Type of proton	Chemical shift in region of
R—CH_3	0.9
—O—CH_2—R	4.0
—O—H	5.0

NMR spectrum of ethanol

HIGH-RESOLUTION NMR SPECTRUM

Very accurate control of the magnetic field will allow a more detailed spectrum to be obtained. Single peaks may then be seen as 'double', 'triple', 'quadruple' or more peaks very close together. This is due to a 'local field' effect and is often referred to as 'spin–spin splitting'. In effect the nature (i.e. multiplicity) of a peak yields information about the neighbouring proton environment.

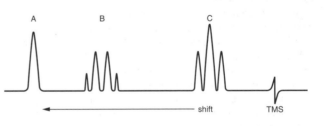

shift   TMS

A has a single peak; as $1 - 1 = 0$ there are no protons on the nearest neighbour atom.

B has a quartet; $4 - 1 = 3$ so the neighbouring carbon has three protons.

C has a triplet; $3 - 1 = 2$ so there are 2 protons on its neighbouring carbon.

NUCLEAR MAGNETIC RESONANCE SPECTROSCOPY

IS ALSO A POWERFUL ANALYTICAL TOOL

Features of molecular spectra

Molecules that are free to move have three types of movement open to them: (a) vibration, (b) rotation, (c) translation.

The explanation for the production of emission spectra involves the notion of quantised energy. Electrons are excited to a higher energy level by absorbing quanta of energy, and on falling back to ground state these quanta are released.

It follows that if energy for electron transitions is quantised, then so is energy for vibration, rotation, and translation. Energy is related to frequency:

$$E = h \times \nu \quad (h = \text{Planck's constant})$$

so particular frequencies are required for particular electron transitions, and also for particular vibrations, etc.

Vibrational levels are must closer than electronic energy levels, and rotational levels closer still. Every electronic energy level has its associated set of vibrational levels, and every vibrational level its own set of rotational levels. Translational energy levels are sets associated with every rotational level. By now the levels are very close and give the semblance of apparent continuity.

When molecules absorb energy sufficient to promote (excite) electrons to a higher electronic energy level some of the increased vibrational energy is readily lost by molecular collision. This means that the emitted radiation will have a different frequency to the energy absorbed. This phenomenon is referred to as **fluorescence**.

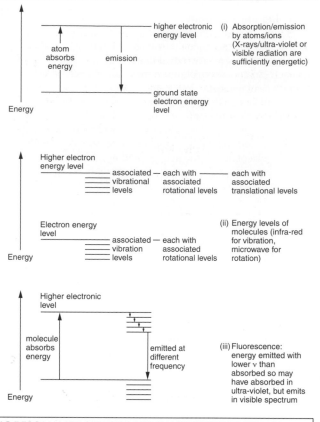

(i) Absorption/emission by atoms/ions (X-rays/ultra-violet or visible radiation are sufficiently energetic)

(ii) Energy levels of molecules (infra-red for vibration, microwave for rotation)

(iii) Fluorescence: energy emitted with lower ν than absorbed so may have absorbed in ultra-violet, but emits in visible spectrum

NB INFRA-RED SPECTROSCOPY AND NUCLEAR MAGNETIC RESONANCE SPECTROSCOPY UTILISE FEATURES OF MOLECULAR SPECTRA

X-ray diffraction

Diffraction is a physical phenomenon that comes about because electromagnetic radiation has a waveform. Separation distances of atoms and ions are of the correct order to cause diffraction of X-rays. The resultant patterns can reveal the sort of information that enables the position of every atom in a complex structure such as a protein to be given. The computations needed to do this have been made possible by the high speed and versatility of modern computers.

Amongst the information gained using X-ray diffraction:

* arrangements of ions in a crystal
* relative positions of atoms in a molecule
* relative positions of molecules in a crystal
* distances between adjacent nuclei (hence atomic size; atomic radii; bond length; bond angle)
* distribution of electrons
* differentiation between stereoisomers

Fluorescence of X-rays gives a powerful analytical tool.

X-ray diffraction is consequent upon electron concentration, and analysis of diffraction patterns makes it possible to draw up **electron-density maps**.

These give direct evidence for the formation of ions (see p. 32); the metallic bond (see p. 31); and delocalisation (see p. 134 for map of benzene).

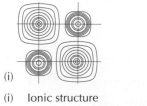

(i)

(ii)

(i)   Ionic structure

(ii)   Metallic structure

(iii)   Covalent compound

Arrangements and separation distances clearly shown. Density contours allow for identification.

(iii)

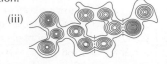

Atomic structure:
the nuclear atom

Nucleus
This contains protons and neutrons, called collectively **nucleons**.

The **mass number** gives the number of nucleons, that is the number of protons + neutrons.

Isotopes are atoms with the *same atomic number* but *different mass numbers*. All the atoms of an element have the same atomic number and it is this that makes them all atoms of a particular element.

The masses of atoms of particular isotopes, called the **relative isotopic mass**, are expressed on a relative scale on which the mass of an atom of the isotope carbon-12 has 12 units exactly.

Relative atomic mass

All elements exist in several isotopic forms and it is useful to have an average value for the masses of the atoms of each element. This is called the **relative atomic mass** and is defined as the weighted mean of the masses of the naturally occurring isotopes of the element expressed on the carbon-12 scale. These are found using a mass spectrometer (see page 22).

The **atomic number** gives the number of protons in the nucleus. It also gives the number of electrons in the neutral atom and the position of the element in the periodic table.

Isotope	Relative isotopic mass	Relative abundance in natural chlorine
^{35}Cl	35	75%
^{37}Cl	37	25%

$$\text{Relative atomic mass} = \frac{(75 \times 35) + (25 \times 37)}{100} = 35.5$$

RADIOACTIVITY

Some isotopes are stable, but others, often with uneven numbers of protons and/or neutrons are unstable. This instability increases with atomic number, resulting from the growing repulsion between increasing numbers of protons. When an unstable isotope **decays** it gives off radiation known as **radioactivity**. This can be in one of three forms as this table shows.

Name of radiation	Made of	Behaviour in electric field	Penetrating power
alpha	helium nuclei	deflected slightly	stopped by paper
beta	electrons	deflected a lot in other direction	stopped by mm of lead
gamma	electromagnetic radiation	similar to X-rays	penetrates cm of lead

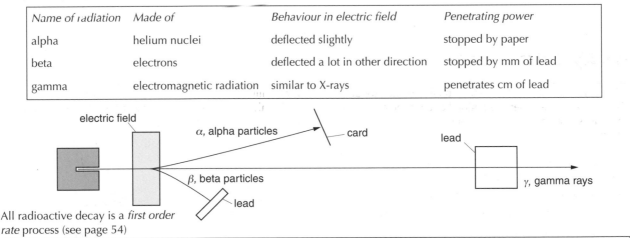

All radioactive decay is a *first order rate* process (see page 54)

Uses of isotopes

Isotopes are often used as **tracers**. Very small traces of an isotope can be detected and hence followed through a process.

- biology e.g. ^{32}P to study nutrient uptake in plants
- industry e.g. ^{57}Fe to study wear and lubrication in engines
- geography e.g. ^{57}Fe to study river flow
- medicine e.g. ^{131}I to study thyroid function
- generating power e.g. ^{235}U in fission reactors; ^{3}H in fusion
- archaeology e.g. ^{14}C in carbon dating

The last example is well known. Nitrogen high in the atmosphere is converted into carbon-14 by cosmic rays coming from space.

The amount of this carbon-14 relative to carbon-12 in the atmosphere was constant until the industrial revolution when the burning of fossil fuels began to dilute it. This means that all carbon-containing objects such as wood and paper started with the same relative amount of carbon-14 as there was in the atmosphere. However, carbon-14 decays back into nitrogen with a half-life of 5568 years, so by measuring the amount of carbon-14 in an object it can be dated to within 200 years. Objects like the Dead Sea scrolls and the Turin Shroud have been dated in this way.

Electronic energy levels, orbitals, and shells: how electrons are arranged in atoms

ENERGY LEVELS

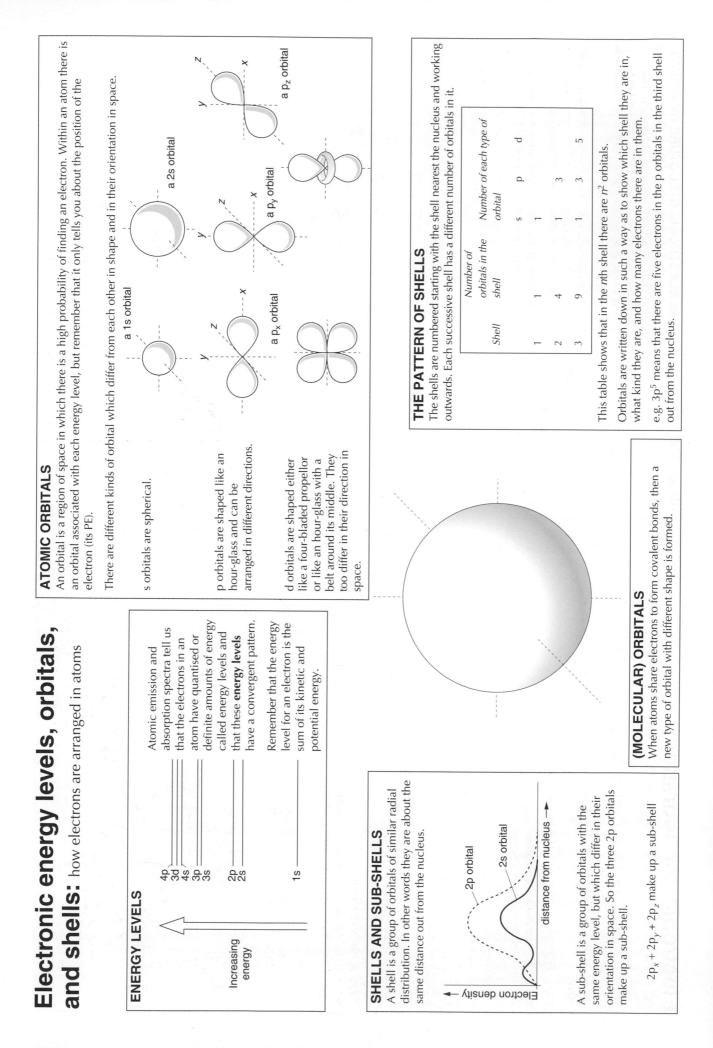

Increasing energy →

4p
3d
4s
3p
3s

2p
2s

1s

Atomic emission and absorption spectra tell us that the electrons in an atom have quantised or definite amounts of energy called energy levels and that these **energy levels** have a convergent pattern.

Remember that the energy level for an electron is the sum of its kinetic and potential energy.

SHELLS AND SUB-SHELLS

A shell is a group of orbitals of similar radial distribution. In other words they are about the same distance out from the nucleus.

Electron density ↑

2p orbital

2s orbital

distance from nucleus →

A sub-shell is a group of orbitals with the same energy level, but which differ in their orientation in space. So the three 2p orbitals make up a sub-shell.

$2p_x + 2p_y + 2p_z$ make up a sub-shell

ATOMIC ORBITALS

An orbital is a region of space in which there is a high probability of finding an electron. Within an atom there is an orbital associated with each energy level, but remember that it only tells you about the position of the electron (its PE).

There are different kinds of orbital which differ from each other in shape and in their orientation in space.

s orbitals are spherical.

a 1s orbital

a 2s orbital

a p_z orbital

a p_y orbital

a p_x orbital

p orbitals are shaped like an hour-glass and can be arranged in different directions.

d orbitals are shaped either like a four-bladed propellor or like an hour-glass with a belt around its middle. They too differ in their direction in space.

THE PATTERN OF SHELLS

The shells are numbered starting with the shell nearest the nucleus and working outwards. Each successive shell has a different number of orbitals in it.

Shell	Number of orbitals in the shell	Number of each type of orbital		
		s	p	d
1	1	1		
2	4	1	3	
3	9	1	3	5

This table shows that in the nth shell there are n^2 orbitals.

Orbitals are written down in such a way as to show which shell they are in, what kind they are, and how many electrons there are in them.

e.g. $3p^5$ means that there are five electrons in the p orbitals in the third shell out from the nucleus.

(MOLECULAR) ORBITALS

When atoms share electrons to form covalent bonds, then a new type of orbital with different shape is formed.

Electronic configurations

represent the electronic structures of atoms

EVIDENCE FOR THE ELECTRONIC STRUCTURE OF ATOMS

A measure of how well an atom can hold its electrons is given by the **ionisation energy**. This is the energy change when a mole of electrons is removed from a mole of particles in the gas phase.

e.g. $X(g) \rightarrow X^+(g) + e^-$ ΔE = ionisation energy

Successive ionisation energies

Successive electrons can be stripped off an atom one after the other until only the nucleus is left. If these successive ionisation energies are plotted (usually on a log scale because the values get so big) against the number of electrons removed, a graph is produced which clearly shows the shell structure of the atom. The graph to the right shows this for potassium.

First ionisation energies

If the first ionisation energies for each element are plotted against atomic number, then once again the shell structure is revealed, but this time in a different way.

RULES FOR WORKING OUT THE ELECTRONIC STRUCTURE OF ATOMS

1. Fill up the orbitals starting with those of lowest energy (nearest the nucleus) and working outwards. This *building up* process is sometimes given the German name the *aufbau principle*. The pattern on the left will help you remember the order.

   An atom whose electrons are in the orbitals of lowest available energy is in its *ground state*. Most atoms are in their ground state at room temperature.

2. Each orbital can have a maximum of two electrons in it (this is known as the *Pauli principle*).

3. When you are filling a sub-shell, half fill each orbital before completely filling any one (known as the *Hund principle*).

Examples

Chlorine: atomic number = 17, so 17 electrons to be placed in orbitals: $1s^2 2s^2 2p^6 3s^2 3p^5$

1st shell   2nd shell   3rd shell

add these to get total no. of electrons

Manganese: atomic number = 25: $1s^2 2s^2 2p^6 3s^2 3p^6 4s^1 3d^5$

4s and 3d orbitals half full

these electrons are higher in energy than the 4s, but nearer the nucleus

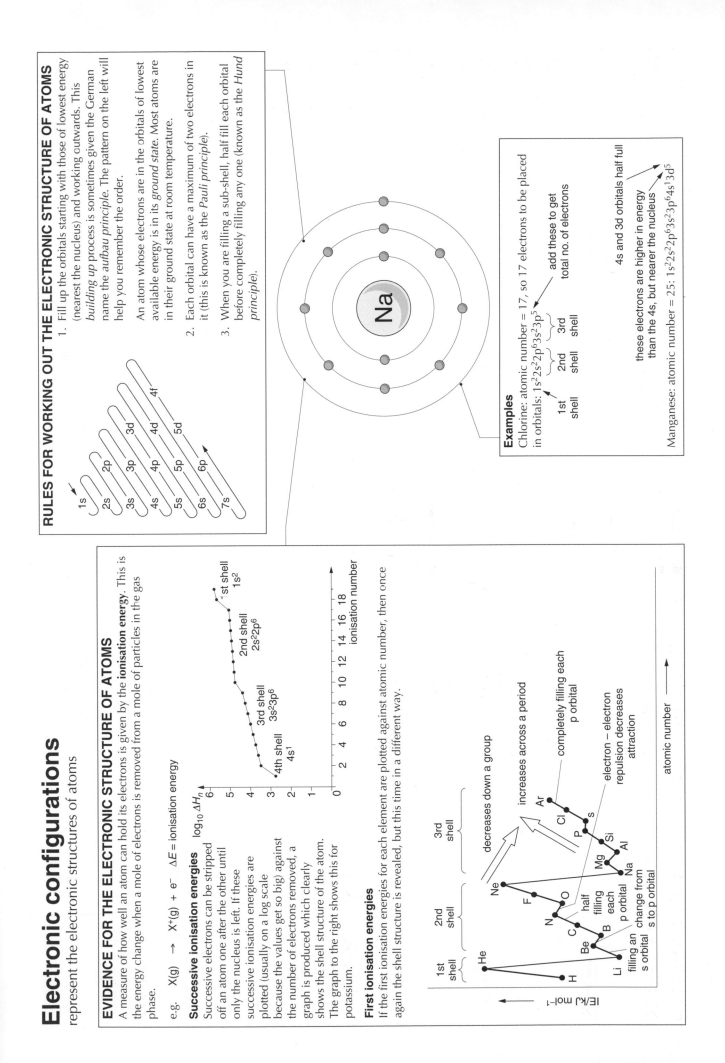

Periodic trends in atomic structure

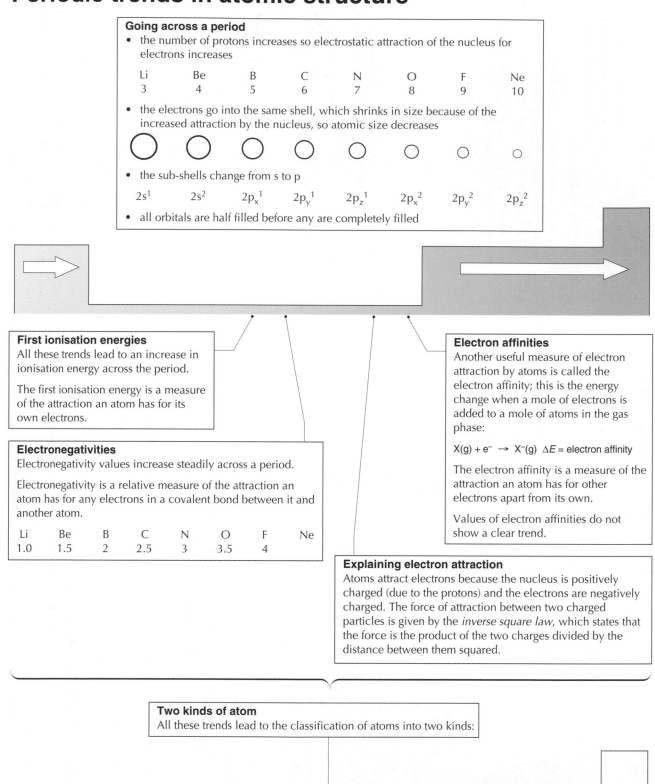

Going across a period

- the number of protons increases so electrostatic attraction of the nucleus for electrons increases

Li	Be	B	C	N	O	F	Ne
3	4	5	6	7	8	9	10

- the electrons go into the same shell, which shrinks in size because of the increased attraction by the nucleus, so atomic size decreases

- the sub-shells change from s to p

$2s^1$ $2s^2$ $2p_x^1$ $2p_y^1$ $2p_z^1$ $2p_x^2$ $2p_y^2$ $2p_z^2$

- all orbitals are half filled before any are completely filled

First ionisation energies

All these trends lead to an increase in ionisation energy across the period.

The first ionisation energy is a measure of the attraction an atom has for its own electrons.

Electronegativities

Electronegativity values increase steadily across a period.

Electronegativity is a relative measure of the attraction an atom has for any electrons in a covalent bond between it and another atom.

Li	Be	B	C	N	O	F	Ne
1.0	1.5	2	2.5	3	3.5	4	

Electron affinities

Another useful measure of electron attraction by atoms is called the electron affinity; this is the energy change when a mole of electrons is added to a mole of atoms in the gas phase:

$X(g) + e^- \rightarrow X^-(g)$ ΔE = electron affinity

The electron affinity is a measure of the attraction an atom has for other electrons apart from its own.

Values of electron affinities do not show a clear trend.

Explaining electron attraction

Atoms attract electrons because the nucleus is positively charged (due to the protons) and the electrons are negatively charged. The force of attraction between two charged particles is given by the *inverse square law*, which states that the force is the product of the two charges divided by the distance between them squared.

Two kinds of atom

All these trends lead to the classification of atoms into two kinds:

more non-metallic

Non-metal atoms: relatively small atoms with strong attraction for electrons

Metal atoms: relatively large atoms with weak attraction for electrons

more metallic

Metallic bonding:
how metal atoms bond together

PROPERTIES OF METALS

1. Conduction

The metal lattice has a very large number of free, *delocalised* outer electrons in it. When a potential gradient is applied, these electrons can move towards the positive end of the gradient carrying charge.

$+$

$-$

2. Ductility

Metals can be bent and reshaped without snapping. The property of bending under tension is called *ductility*; bending under pressure is called *malleability*.

This can happen in metals because the close-packed layers can slide over each other without breaking more bonds than are made.

Impurities added to the metal disturb the lattice and so make the metal less ductile. This is why alloys are harder than the pure metals they are made from.

A **metallic bond** is the electrostatic force of attraction that two neighbouring nuclei have for the delocalised electrons between them. Both ions attract the delocalised electrons between them leading to metallic bonding.

$+$ $+$

Most metals exist in *close-packed lattices of ions surrounded by delocalised outer electrons.*

delocalised electrons

nucleus and inner shells

The close-packed layers can be vertically stacked together in an ab ab ab sequence, which means that every other layer is vertically lined up, or in an abc abc sequence, when every third layer is vertically lined up. The first produces a lattice called *hexagonal close packed* (*h.c.p.*); magnesium and zinc are examples, while the second is called *face centred cubic* (*f.c.c.*); aluminium and copper are examples. A minority of metals are cubically packed in which case the lattice is called *body centred cubic* (*b.c.c.*); iron and chromium are examples.

METALLIC LATTICE ENERGIES

It is not correct to talk about bond strength in metallic lattices. Instead we refer to the *lattice energy*.

This is the energy needed to break up one mole of atoms in the lattice into separate atoms:

$$M(s) \rightarrow M(g)$$

The figures below show that the factors affecting the lattice energy of a metal are *size of the cations*, the *charge on the cation*, and *the kind of lattice*

Gp 1		**Gp 2**		**Gp 3**	
Li	159	Be	314	Al	314
Na	106	Mg	151		

Going down a group the lattice energy decreases as the size of the ions increases.

Going across the period from group 1 to 2 to 3 the lattice energy increases as the charge on the ion increases.

So high lattice energies result from small highly charged ions.

Ionic bonding: how metal atoms bond to non-metal atoms

PROPERTIES OF IONIC COMPOUNDS

1. Conduction

In the solid state the ions are held tightly in the lattice and cannot move to carry their charges: so in the solid state ionic substances are insulators.

When they are molten or dissolved, the ions can move and carry their charges through the liquid

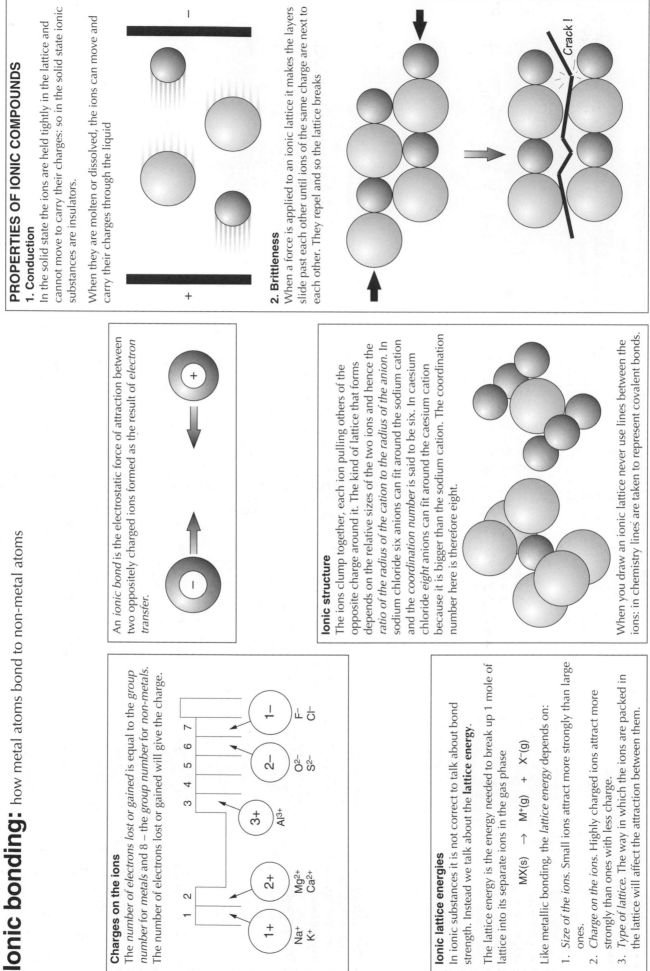

2. Brittleness

When a force is applied to an ionic lattice it makes the layers slide past each other until ions of the same charge are next to each other. They repel and so the lattice breaks

Crack !

An *ionic bond* is the electrostatic force of attraction between two oppositely charged ions formed as the result of *electron transfer*.

Charges on the ions

The *number of electrons lost or gained* is equal to the *group number* for metals and 8 – the group number for non-metals.

The number of electrons lost or gained will give the charge.

1   2   3   4   5   6   7

1+   2+   3+   2–   1–

Na⁺   Mg²⁺   Al³⁺   O²⁻   F⁻
K⁺   Ca²⁺   S²⁻   Cl⁻

Ionic structure

The ions clump together, each ion pulling others of the opposite charge around it. The kind of lattice that forms depends on the relative sizes of the two ions and hence the *ratio of the radius of the cation to the radius of the anion*. In sodium chloride six anions can fit around the sodium cation and the *coordination number* is said to be six. In caesium chloride *eight* anions can fit around the caesium cation because it is bigger than the sodium cation. The coordination number here is therefore eight.

When you draw an ionic lattice never use lines between the ions: in chemistry lines are taken to represent covalent bonds.

Ionic lattice energies

In ionic substances it is not correct to talk about bond strength. Instead we talk about the **lattice energy**.

The lattice energy is the energy needed to break up 1 mole of lattice into its separate ions in the gas phase

$$MX(s) \rightarrow M^+(g) + X^-(g)$$

Like metallic bonding, the *lattice energy* depends on:

1. *Size of the ions.* Small ions attract more strongly than large ones.
2. *Charge on the ions.* Highly charged ions attract more strongly than ones with less charge.
3. *Type of lattice.* The way in which the ions are packed in the lattice will affect the attraction between them.

Covalent bonding
is the bonding between two non-metallic atoms

COVALENCY
The number of bonds formed is known as the **covalency** or valency for short.

To form a bond an atom usually needs an electron to put into the bond and a space in an orbital to accept the electron from the other atom. For atoms of elements in the second period, the number of electrons in the bonding (outer) shell is limited to eight. The table shows you how the number of bonds formed by an atom is related to the position of the element in the periodic table.

Element	B	C	N	O	F	Ne
No. of electrons in outer shell	3	4	5	6	7	8
No. of spaces in outer shell	5	4	3	2	1	0
No. of bonds made	3	4	3	2	1	0
Examples						

lone pairs

THE COVALENT BOND
A covalent bond is the electrostatic force of attraction that two neighbouring nuclei have for a localised pair of electrons *shared* between them.

Drawn like this H⚬×H to show electronic structure or like this H–H to show the bond

NON-BONDING OR LONE PAIRS
Notice that in the first two examples there are only shared pairs, but in the last four there are also pairs of electrons which are not shared, but belong only to the central atoms. Pairs of electrons like this, which are under the control of only one atom, are called **non-bonding** or **lone pairs.**

DATIVE COVALENCY AND COORDINATE BONDS
Sometimes both the electrons in a covalent bond come from only one of the atoms. This is called dative covalency and the bond is called a coordinate bond. Once the bond has formed it is identical to any other covalent bond. It does not matter which atom the electrons came from.

e.g.

ammonium ion, NH₄⁺

coordinate bond

electronic structure

bonding

hydroxonium ion, H₃O⁺

electronic structure

bonding

TWO KINDS OF COVALENT BOND
Covalent bonds form when the orbitals of two neighbouring atoms overlap so that both nuclei attract the pairs of electrons between them. This can happen in two different ways making two different kinds of bond:

Sigma, σ, bonds

When the orbitals from two atoms overlap along the line drawn through the two nuclei, a sigma bond forms.

Pi, π, bonds

Sometimes, after a sigma bond has formed between two atoms, the p orbitals of the two atoms also overlap above and below the line drawn through the two nuclei and another bond forms. This is called a pi bond and is made of two regions of electron density.

e.g. here two p orbitals overlap after a sigma bond has formed.

e.g. two s orbitals, an s and a p orbital, or two p orbitals can overlap.

σ bonds

These are examples of molecular orbitals.

π bond

σ bond

Overlap above and below line of centres

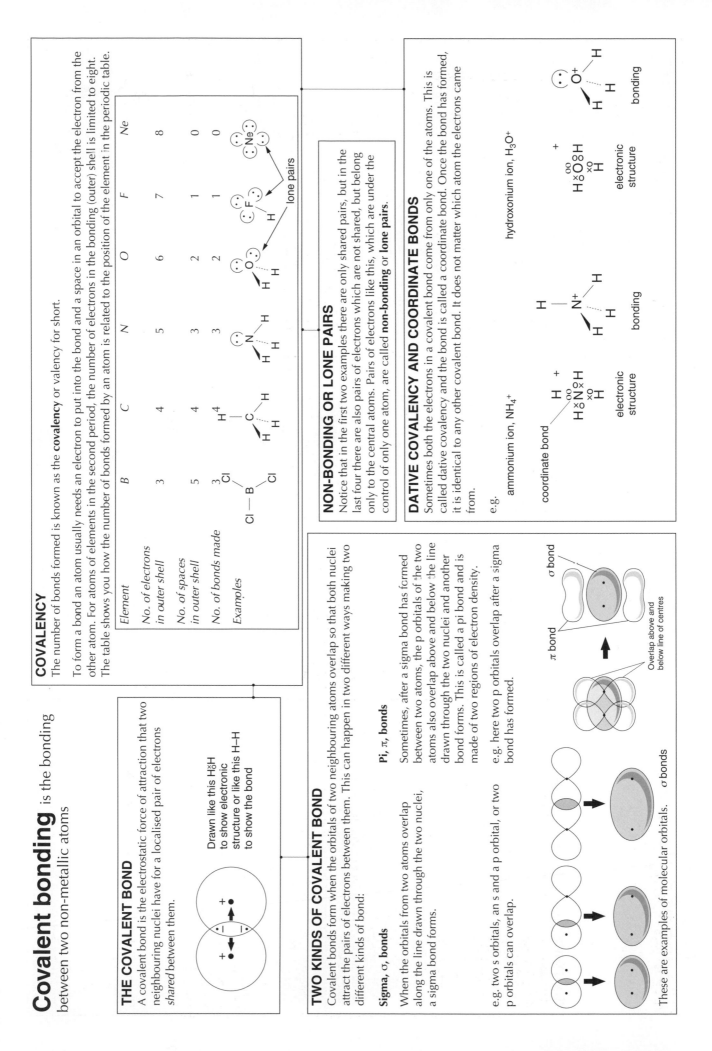

Shapes of molecules I

ELECTRON PAIR REPULSION

The shape of a molecule can be worked out using the following rules about electron pairs:

(i) the pairs of electrons in a molecule (both shared and lone pairs) behave as though they repel each other and so move as far apart as possible;

(ii) lone pairs repel more than bonding pairs.

WORKING OUT THE SHAPES OF PARTICLES CONTAINING ONLY SINGLE BONDS

The number of electron pairs in molecules containing only single bonds is found by following these steps:

1. Decide which atom is at the centre of the molecule (usually the atom of which there is only one).

2. Look up which periodic group it is in: this gives the number of outer electrons it has.

3. Add an electron if the particle is negatively charged; subtract an electron if the particle is positively charged.

4. Add one electron for each atom joined to the central atom.

5. Divide by two to get the number of pairs.

6. This gives the shape of the electron pairs. Now see if there are the same number of bonded atoms; if not, the extra pairs are lone pairs.

7. Now name the shape of the molecule which is defined only by the arrangement of nuclei.

EXAMPLES OF WORKING OUT THE SHAPES OF MOLECULES

Steps	H_2O	NH_3	PCl_5	SCl_4	PCl_4^+	ICl_4^-
1	O	N	P	S	P	I
2	6	5	5	6	5	7
3	0	0	0	0	−1	+1
4	2	3	5	4	4	4
5	$\frac{8}{2}=4$	$\frac{8}{2}=4$	$\frac{10}{2}=5$	$\frac{10}{2}=5$	$\frac{8}{2}=5$	$\frac{12}{2}=6$
6						
7	bent	trigonal pyramidal	trigonal bipyramidal	distorted tetrahedron	tetrahedral	square planar

So, the number of electron pairs decides the shape as the table shows:

No. of charge clouds	Shape	Name of shape
2		linear
3		trigonal planar
4		tetrahedral
5		trigonal bipyramidal
6		octahedral

Shapes of molecules II

EXPLAINING DIFFERENT BOND ANGLES

The exact angles between bonds can be found using X-ray diffraction. These angles are evidence for the idea that lone pairs repel more than bonding pairs. For example, the angles in ammonia and water are less than those in methane, as shown below.

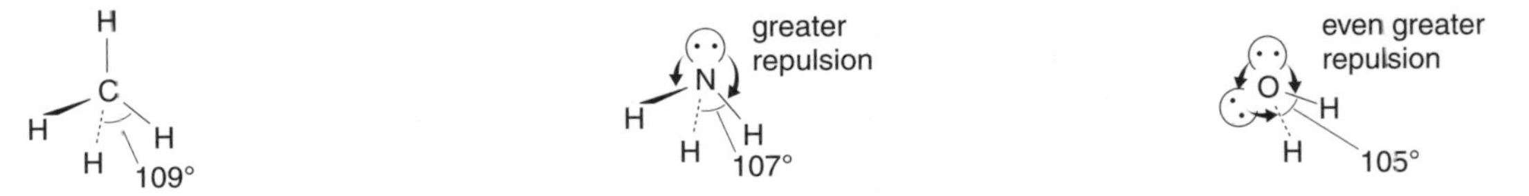

even greater repulsion — H₂O 105°

greater repulsion — N, H 107°

C, H 109°

WORKING OUT THE SHAPES OF OXO-COMPOUNDS

Many molecules and ions contain oxygen. This oxygen is: sometimes double bonded and sometimes single bonded, in which case the oxygen atom carries a minus charge.

The shape of these particles can be worked out by following these steps:

1. For each minus charge on the particle there is one single bonded oxygen atom with a minus charge.

2. All the other oxygens are double bonded to the central atom. This double bond counts as one electron charge cloud when working out shapes.

3. So draw the two types of oxygen around the central atom and check the number of electrons it is using in bonds to the oxygens.

4. If there are any unused lone pairs around the central atom put them in, then choose the shape for the number of electron charge clouds around the central atom.

DELOCALISATION

The structure of the carbonate anion shows two single bonds and one double bond. In fact experimental data show that all the bonds in the carbonate ion are the same length and all the bond angles are the same. They are longer than a C=O bond but shorter than a C–O bond. This shows that the bonding model we have used is wrong and in fact the π bond has been *delocalised* throughout the molecule. The charge has also been spread over the whole molecule rather than being concentrated on only two of the oxygens.

Delocalisation occurs in most oxo-anions (e.g. nitrate, sulphate, ethanoate, manganate, etc.) and in some organic molecules such as benzene (see page 134).

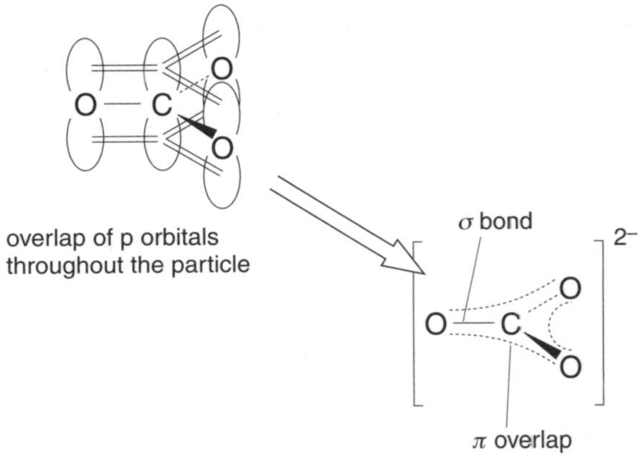

overlap of p orbitals throughout the particle

σ bond   π overlap

THE EXAMPLES BELOW SHOW THE PROCESS IN ACTION:

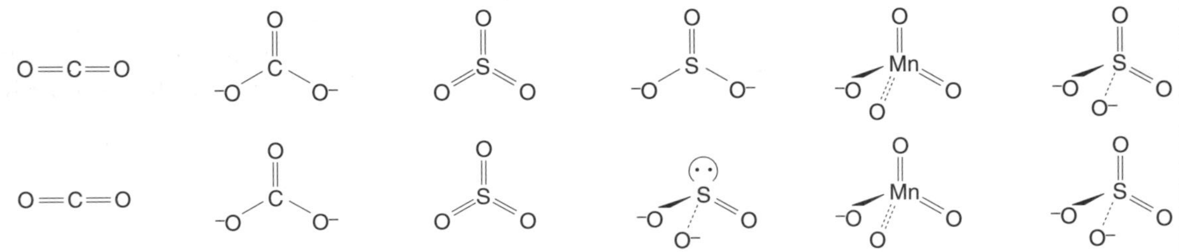

Steps	CO_2	CO_3^{2-}	SO_3	SO_3^{2-}	MnO_4^-	SO_4^{2-}
1	None	2—O⁻	None	2—O⁻	1—O⁻	2—O⁻
2	2=O	1=O	3=O	1=O	3=O	2=O
3	O=C=O	planar trigonal	planar trigonal	trigonal pyramidal		
4	O=C=O linear	planar trigonal	planar trigonal	trigonal pyramidal	tetrahedral	tetrahedral

Charge separation and dipole moments

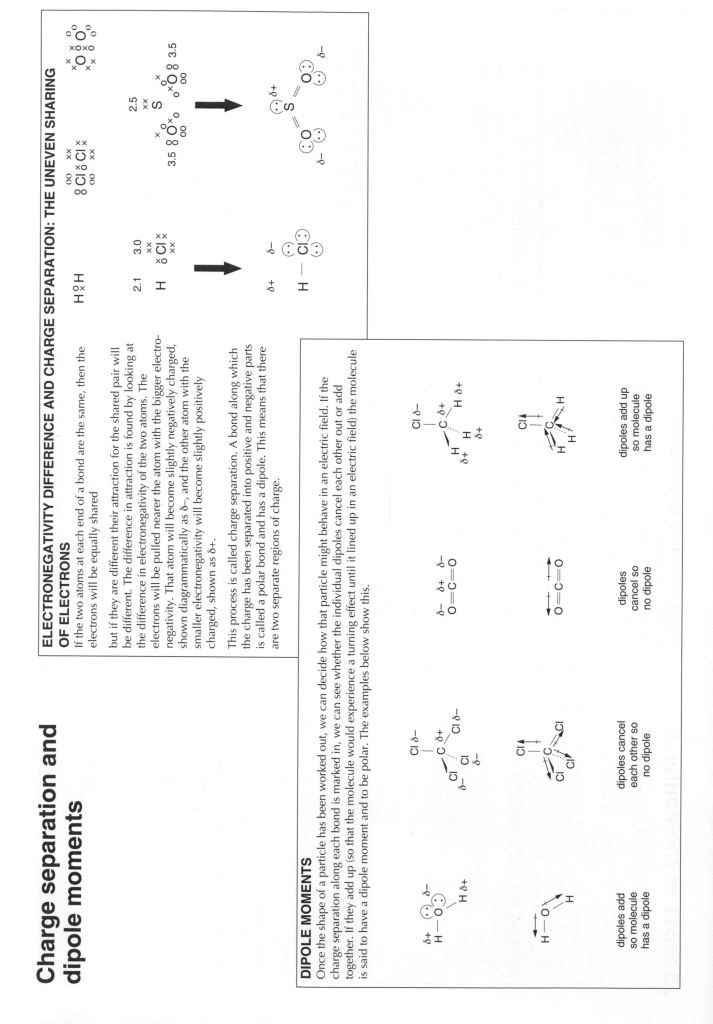

ELECTRONEGATIVITY DIFFERENCE AND CHARGE SEPARATION: THE UNEVEN SHARING OF ELECTRONS

If the two atoms at each end of a bond are the same, then the electrons will be equally shared

but if they are different their attraction for the shared pair will be different. The difference in attraction is found by looking at the difference in electronegativity of the two atoms. The electrons will be pulled nearer the atom with the bigger electronegativity. That atom will become slightly negatively charged, shown diagrammatically as δ−, and the other atom with the smaller electronegativity will become slightly positively charged, shown as δ+.

This process is called charge separation. A bond along which the charge has been separated into positive and negative parts is called a polar bond and has a dipole. This means that there are two separate regions of charge.

DIPOLE MOMENTS

Once the shape of a particle has been worked out, we can decide how that particle might behave in an electric field. If the charge separation along each bond is marked in, we can see whether the individual dipoles cancel each other out or add together. If they add up (so that the molecule would experience a turning effect until it lined up in an electric field) the molecule is said to have a dipole moment and to be polar. The examples below show this.

Forces in covalently bonded substances

FORCES INSIDE THE MOLECULES

Single covalent bonds

The strength of a covalent bond – the bond dissociation energy – depends on how strongly the two nuclei attract the bonding pairs between them. Going down any periodic group, the bonds between the atoms of the group and another atom will get weaker because the bonding pair gets further away from the nuclei.

Bond	BDE kJ mol^{-1}		Bond length nm
H–F	562		0.092
H–Cl	431		0.128
H–Br	366		0.141
H–I	299		0.160

Multiple bonds

When there are double bonds between atoms the second (π) bond is not as strong as the first (σ) bond. This is because the overlap of the orbitals is not so good and the shared pairs are further from the nuclei than those in the sigma bond.

C–C	348 kJ mol^{-1}
C=C	612 kJ mol^{-1} — not twice as much so a different kind of bond

FORCES BETWEEN MOLECULES

Van der Waals forces

The electrons in any particle can at any moment be unevenly spread so that one side of the particle is slightly positive while the other is negative. This is called an **instantaneous dipole**. The more electrons there are in a particle the more likely it is that a dipole exists.

An instantaneous dipole in one particle can induce another dipole in any neighbouring particle. The two particles will then attract each other, an instantaneous dipole-induced attraction.

These weak forces of attraction between all particles are often called **van der Waals forces**. The strength of the van der Waals forces depends on the number of electrons in the particles. This is shown by the correlation between the number of electrons and the boiling point.

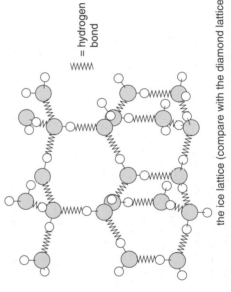

the attraction between them is a van der Waals force

Molecule	No. of electrons	b.p. °C
H$_2$	2	−253
N$_2$	14	−196
Cl$_2$	34	−35
I$_2$	126	+185

Hydrogen bonds

When hydrogen is bonded to nitrogen, oxygen, or fluorine, a particularly strong dipole is made. This happens for two reasons. First, nitrogen, oxygen, and fluorine are very electronegative so they pull the bonding pair between them and hydrogen well away from the hydrogen, making it quite positive. Second, nitrogen, oxygen, and fluorine atoms have lone pairs of electrons which are attracted to the positive hydrogen atom. This force of attraction between an electron-deficient hydrogen bonded to nitrogen, oxygen, or fluorine and the lone pair of a neighbouring nitrogen, oxygen, or fluorine atom is called a hydrogen bond.

$\wedge\wedge\wedge\wedge$ = hydrogen bond

the ice lattice (compare with the diamond lattice)

Dipole–dipole forces

Particles with permanent dipoles attract each other more strongly than those with only van der Waals forces between them. Even molecules like carbon dioxide, which do not have dipole moments, have strong dipole–dipole forces between the oppositely charged regions of neighbouring molecules.

$$\delta{-}O = C = O \delta{-}$$
$$\delta{+}C$$
$$\delta{-}O = C = O \delta{+} \quad C\delta{+}$$
$$\delta{-}O$$

dipole/dipole forces

The relative strength of the different kinds of bonds

The data book shows that covalent bonds are 100s of kJ strong. Hydrogen bonds are about ten times weaker, while van der Waals forces are ten times weaker still. The van der Waals forces which are present in all systems only become obvious if there are no stronger bonds present to mask their effect. So as a rough rule:

covalent bonds ≈ 10 × hydrogen bonds ≈ 100 × van der Waals forces

The structure of solids: relating

the properties of solids to the types of lattice

Bonding and structure

These two words are often used together and are therefore confused with each other.

The word bonding refers to the different kinds of attractive force between the particles.

The word structure refers to the arrangement of particles.

Lattice

A lattice is an ordered arrangement of particles held together by some form of bonding. This bonding may be strong in three, two, or only one dimension and this leads to three kinds of lattice.

STRONG BONDING IN ONE DIMENSION

Ionic

e.g. copper (II) chloride

Covalent

e.g. protein, starch, or hydrocarbon chain

Part of a starch molecule: a carbohydrate

Part of a protein chain

Part of a hydrocarbon chain: polythene

STRONG BONDING IN TWO DIMENSIONS

Ionic

e.g. aluminium chloride

weak bonding between layers

sandwich layers with strong bonding between small aluminium ions and large chloride ions

Covalent

e.g. graphite

layers of delocalised electrons between the layers

STRONG BONDING IN THREE DIMENSIONS

Ionic

e.g. sodium chloride

Covalent

e.g. diamond and quartz

diamond lattice

silicon

oxygen

silicon dioxide lattice

Bonding summary

- Decide on the kind of bonding by looking at the kinds of atoms

between metallic atoms → **METALLIC BONDING**	between metallic and non-metallic atoms → **IONIC BONDING**	between non-metallic atoms → **COVALENT BONDING**
here electrons are *shared* and *delocalised*	here electrons are transferred: outer shell electrons are *lost* by *metal* atoms and *gained* by *non-metal* atoms	here electrons are *shared* and *localised*: along the line of atomic centres → σ bond; above and below the line of atomic centres → π bond

- Think about the electronic structure of the actual atoms involved. From this you can find:

| the *charges* on the ions in the metal lattice:

this is the same as the *group number* | the *charges* on the ions:

for metals it is the same as the *group number*,

for non-metals the charge equals

8 – *group number* | the number of electrons shared and in lone pairs:

so the number of σ and π pairs |

- Assess the bonding model

| | Decide if there is any *covalent character*.

- Is the cation small and highly charged?
- Is the anion large and highly charged?

If so there will be some covalent character. | Look at the *electronegativities* of the two atoms:

- if they are very different there will be some *charge separation* along the bond of the two atoms |

- Think about the arrangement of particles

| In a metal lattice the cations tend to be *close packed* with *delocalised outer shell electrons* | Look at the *ratio of radius of the cation to radius of the anion*:

this gives the arrangement of ions in the lattice: | Work out the shape of the molecule using the idea of *electron pair repulsion* which is that pairs of electrons behave as though they repel each other and so get as far apart as possible

Is there charge separation along the bonds in the molecule and do the dipoles caused by this cancel out? Is the molecule symmetrical?
- if it is not it will be *polar*

Is the substance *macromolecular*?

Is the substance *molecular*?
- if so:
- the *number of electrons* indicates the strength of *van der Waals* forces
- *charge separation* indicates strength of *dipole–dipole* forces
- *H bonded to N, O, or F* means *hydrogen bonding*

the m.p. and b.p. will depend on the strength of these bonds; as a general rule van der Waals forces are 10× weaker than hydrogen bonds which are 10× weaker than covalent bonds. |

- Decide how strong the forces are

| *Lattice energy* depends on

- *size*
- *charge* of the cations:

remember that the *d block* metal nuclei are *poorly shielded* and so have higher effective nuclear charge. | *Lattice energy* depends on

- *sizes*
- *charges* of the ions
- and the *kind of lattice* | Covalent bond strength depends on the distance apart of the atomic centres and hence the *sizes of the atoms*. |

Isomerism I: structural isomerism

CLASSIFYING ISOMERS

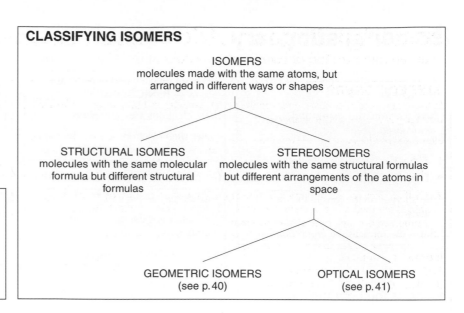

ISOMERS
molecules made with the same atoms, but arranged in different ways or shapes

STRUCTURAL ISOMERS
molecules with the same molecular formula but different structural formulas

STEREOISOMERS
molecules with the same structural formulas but different arrangements of the atoms in space

GEOMETRIC ISOMERS
(see p. 40)

OPTICAL ISOMERS
(see p. 41)

In **structural isomers** the atoms are bonded together in a different order so that although they have the *same molecular formula* their *structural formulas are different*.

Structural isomers are found in both inorganic and organic substances.

Inorganic example:
the chlorides of chromium
These differ depending on whether the chloride ions are free in solution or bonded as ligands to the chromium cation.

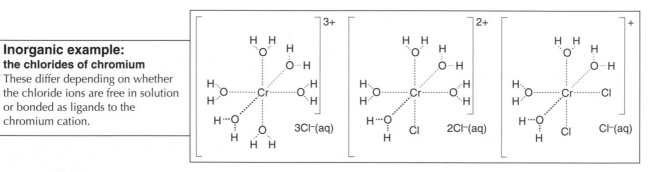

Organic examples
1. Straight and branched chain alkanes:
e.g. there are isomers of pentane, C_5H_{12}

2. Alcohols and ethers:
e.g. ethanol and methoxymethane, C_2H_6O

3. Halogenoalkanes:
e.g. dichloroethane, $C_2H_4Cl_2$

4. Benzene derivatives
e.g. hydroxybenzoic acid

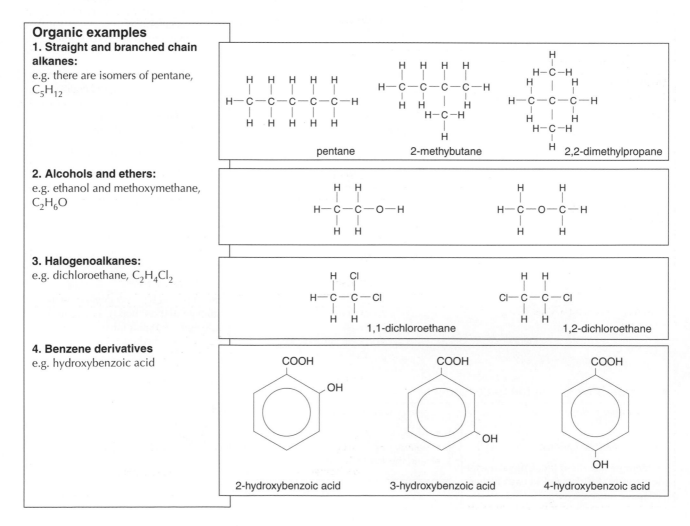

pentane   2-methybutane   2,2-dimethylpropane

1,1-dichloroethane   1,2-dichloroethane

2-hydroxybenzoic acid   3-hydroxybenzoic acid   4-hydroxybenzoic acid

Isomerism II: geometric stereoisomerism
is caused by restricted rotation in a molecule

STEREOISOMERS

In these the molecules have the *same structural formulas*, but they have a ***different arrangement of the atoms in space***.

Stereoisomers happen for two reasons:

1. Either there is restricted rotation in the molecule (caused by a double bond or ring) in which case the isomerism is called **geometric stereoisomerism**.

2. Or there is no plane or point of symmetry in the molecule (four different groups joined tetrahedrally to a central atom is a common case) in which case the isomerism is called **optical stereoisomerism**.

GEOMETRIC STEREOISOMERISM

Look for a double bond with two different groups joined to the atoms at each end of the bond, or a ring structure again with two different groups joined to two neighbouring atoms in the ring.

The two isomers are called *cis* (on the same side) and *trans* (on opposite sides) isomers.

Inorganic examples

1. One of the chlorides of chromium, $[Cr(H_2O)_4Cl_2]^+$, has geometric isomers

2. Complexes with bidentate ligands

Organic examples

1. Isomers based on alkene structures, $C_2H_2Cl_2$

2. Isomers based on oxime structures, CH_3CHNOH

3. Isomers based on aliphatic ring structures

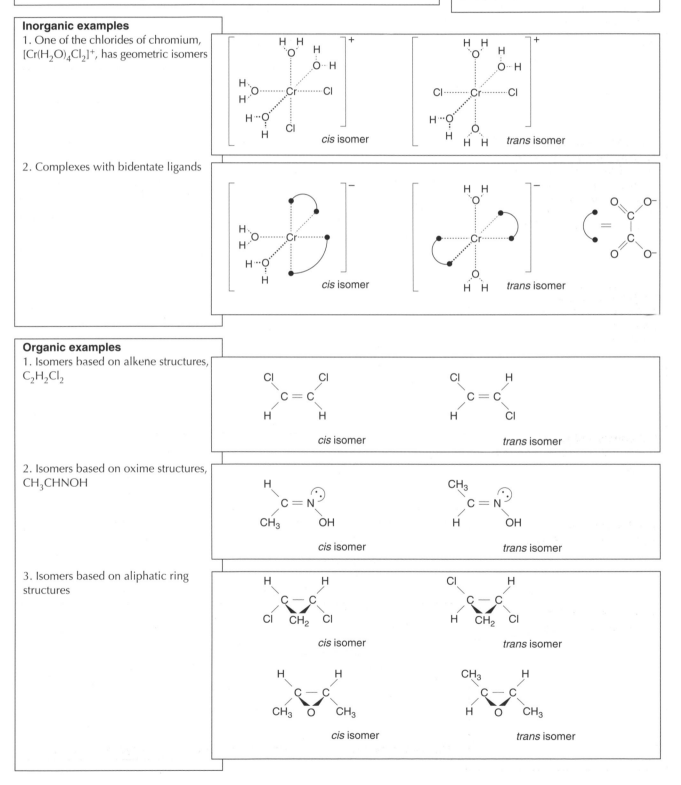

Isomerism III: optical isomerism

Molecules which have no plane or point of symmetry rotate the plane of polarised light

Chirality

All objects can have mirror images, but only certain objects have the property of handedness or chirality. The mirror images of these objects cannot be superimposed on each other because of their handedness.

Gloves, scissors, screws, and propellors are all chiral objects.

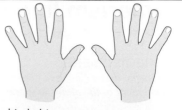

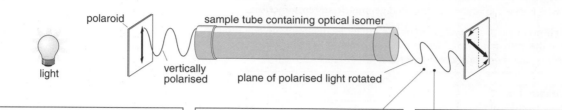

When drawing optical isomers always first draw in a mirror line and then as you draw each part of the first isomer, draw its reflection the other side of the mirror line.

Molecules which do this have the property of **handedness** in the same way that gloves or shoes are right or left handed. They are said to be **chiral**. Most molecules do not have this property of handedness and are **achiral**.

Each form of an optical isomer is called an **enantiomer** or **enantiomorph** and one will rotate the plane of polarised light one way while the other will rotate it in the opposite direction.

Inorganic examples are based on complex ions containing bidentate ligands.

Organic examples usually involve molecules in which four different groups are joined to a carbon atom.

A mixture which contains equal quantities of each isomer is called a **racemic mixture** or **racemate**. It will not affect polarised light because there are equal numbers of molecules rotating the light each way. Some reactions make racemic mixtures. Some reactions — especially in living systems — produce only one optical isomer.

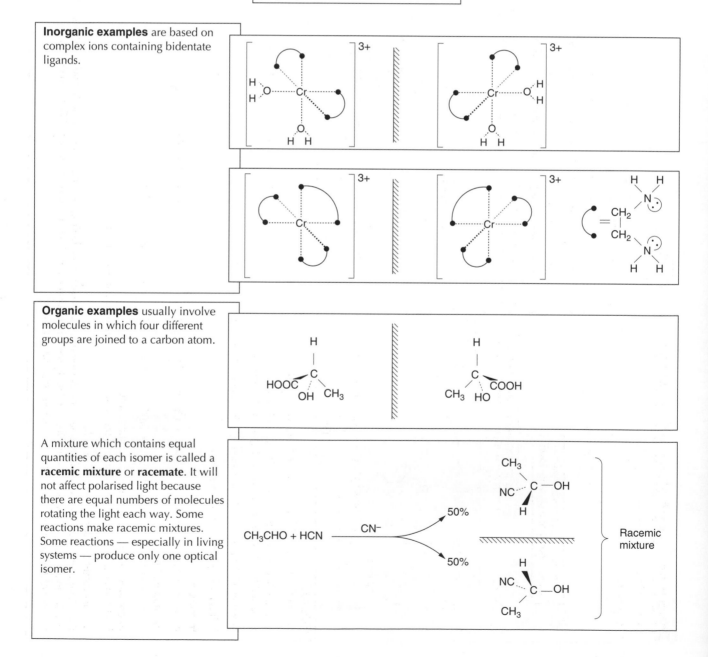

Phases or states of matter:

the physical properties of solids, liquids, and gases

LIQUIDS

The particles in a liquid are fairly well ordered over a short distance, but there is no long range order.

The particles have more kinetic energy than in the solid state and it is this movement of the particles that disrupts the arrangement of the lattice.

The potential energy of the particles is also greater than in solids because they have moved apart slightly.

At room temperature most substances which are liquid are:

- covalently bonded molecular substances with quite strong van der Waals forces (large molecules with lots of electrons) or
- hydrogen bonded liquids such as water and alcohols.

In an **ideal liquid** the behaviour of a particle depends only on the number of other particles around it and not on their identity. Liquid mixtures which behave in this way are said to obey **Raoult's law.**

GASES

The particles in a gas move rapidly and randomly.

The kinetic energy of the particles is very high and all order has been lost.

The particles are far apart and so also have a high potential energy.

Gases are either covalently bonded molecular substances with weak intermolecular forces (small molecules with few electrons) or atomic substances like the noble gases.

To simplify the study of gases we assume that the particles:

- are moving randomly
- do not attract each other
- have no volume
- and have elastic collisions

These assumptions are known as the **postulates of the kinetic theory** and a gas in which the particles behave like this is called an **ideal gas**. In such a gas the kinetic energy of the particles is a measure of their temperature.
For an ideal gas the equation:

$$PV = nRT$$ can be applied

Real gases approach ideal behaviour when the pressure is very low (the particles are far apart and do not attract each other) and/or when the temperature is very high (the particles are moving very fast and not near each other).

Conversely, gases are least ideal at high pressure and low temperature.

CHANGING STATE

gas cooling: particles losing kinetic energy and slowing down

gas condensing: particles losing potential energy and getting closer

liquid cooling: particles losing K.E. and slowing down

liquid freezing: particles losing P.E. and getting closer

solid cooling

freezing complete

freezing begins

condensing complete

condensing begins

Time →

b.p.

f.p./m.p.

T°C

SOLIDS

The particles in a solid are arranged in an ordered **lattice**.

The kinetic energy of the particles is low and they vibrate about their lattice position. As the solid is heated the particles move more and the lattice expands becoming more disordered.

The potential energy of the particles is also low because they are close together.

Solids may be bonded in different ways:

In metals
the lattice energy depends on the charge on the metallic ions, the size of the ions, and the type of lattice.

In ionic solids
the lattice energy depends on the charge on the ions, the size of the ions, and the type of lattice.

In covalently bonded macromolecular solids
the bond energy depends on the size of the atoms and the arrangement of the lattice.

In covalently bonded molecular solids
the lattice energy depends on the forces between the molecules. These can be hydrogen bonds in compounds where hydrogen is bonded to nitrogen, oxygen, or fluorine (e.g. H_2O); dipole forces where there is charge separation (e.g. CO_2); van der Waals forces which depend on the number of electrons (e.g. noble gases).

Phase equilibria: the behaviour of substances as they change state

VAPOUR PRESSURE OF SOLUTIONS

Adding an impurity lowers the vapour pressure of a liquid.

This lowering of vapour pressure is caused by the fact that there are fewer solvent particles on the surface of the liquid so fewer can escape into the vapour phase.

The more impurity that is added, the greater the lowering of the vapour pressure.

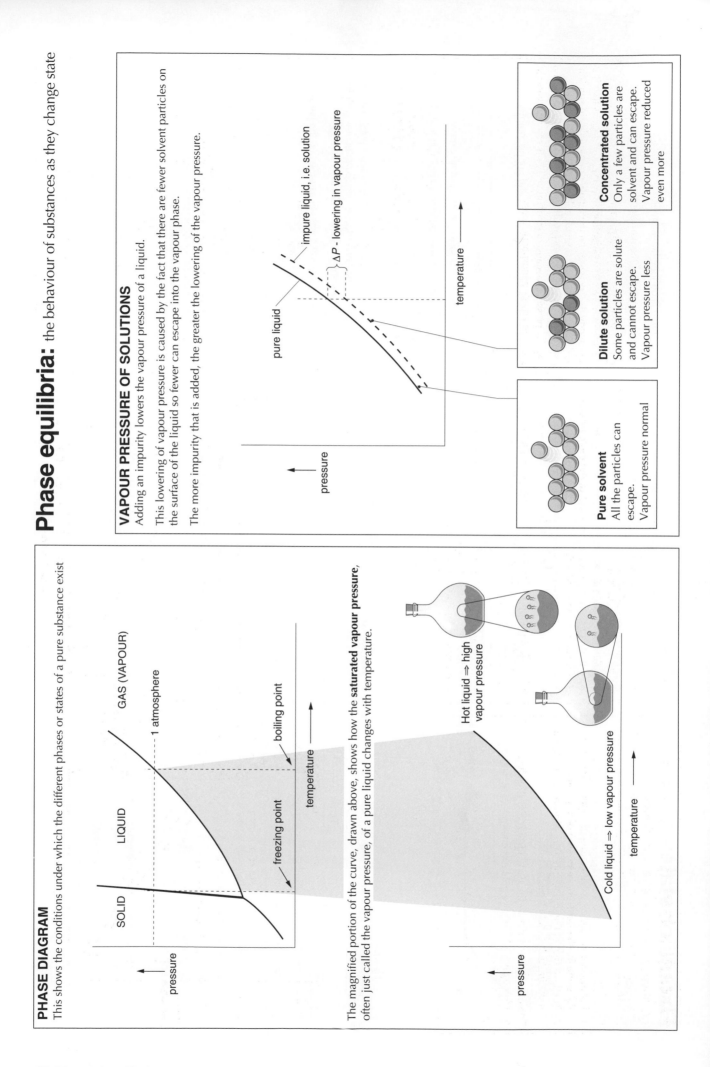

impure liquid, i.e. solution

ΔP - lowering in vapour pressure

pure liquid

pressure

temperature

Pure solvent
All the particles can escape.
Vapour pressure normal

Dilute solution
Some particles are solute and cannot escape.
Vapour pressure less

Concentrated solution
Only a few particles are solvent and can escape.
Vapour pressure reduced even more

PHASE DIAGRAM

This shows the conditions under which the different phases or states of a pure substance exist

pressure

SOLID   LIQUID   GAS (VAPOUR)

1 atmosphere

freezing point

boiling point

temperature

The magnified portion of the curve, drawn above, shows how the **saturated vapour pressure**, often just called the vapour pressure, of a pure liquid changes with temperature.

Hot liquid ⇒ high vapour pressure

Cold liquid ⇒ low vapour pressure

pressure

temperature

Raoult's law

Raoult studied the vapour pressure of solutions and discovered that it does not matter what the solute is, but only how much solute there is dissolved in the solution, or more precisely:

1. For molecular solutes, the lowering of vapour pressure is the same for the same concentration of any solute.
2. The lowering of the vapour pressure is proportional to the mole fraction of solute added.

The mole fraction means the number of moles of solute divided by the total number of moles.

i.e.

$$\text{mole fraction of A} = \frac{\text{no. of moles of A}}{\text{total no. of moles}}$$

Statement of Raoult's law

The relative lowering of the vapour pressure of a pure solvent at constant temperature is equal to the mole fraction of non-volatile solute added.

$$P_x = P_0 m_x$$

where P_x = vapour pressure of the solution, P_0 = vapour pressure of the pure solvent, and m_x = mole fraction of solvent in the solution.

A solution which obeys Raoult's law is called an **ideal solution**. In an ideal solution, the force between the particles is constant and does not depend on their identity.

COLLIGATIVE PROPERTIES

Vapour pressure is a **colligative property**. This means that it is a property that depends on the **number** of particles in solution and not their identity. Another colligative property is the **elevation in boiling point** of the solution. This depends on the vapour pressure, because a solution will boil when its vapour pressure equals atmospheric pressure.

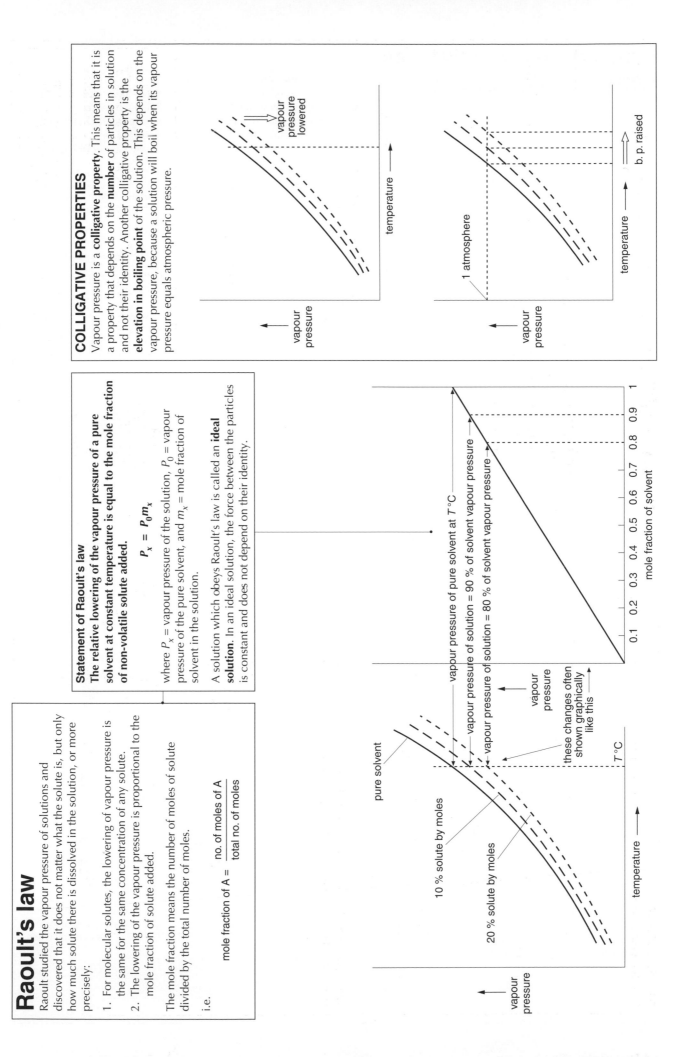

Homogeneous mixtures with volatile solutes

Homogenous mixtures with volatile solutes

In a mixture of two liquids which mix, particles from both liquids escape from the surface creating a partial pressure. If both liquids obey Raoult's law, the graph below is produced. **Dalton's law of partial pressures** states that the total pressure in a mixture is the sum of the partial pressures so a total pressure line can be added to the graph as shown here.

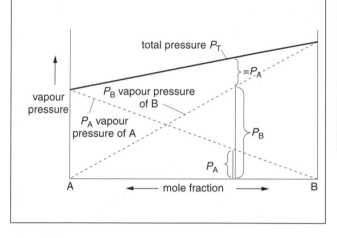

Vapour composition curves

Any liquid mixture will evaporate, but the composition of the vapour above the liquid will be different from the composition of the liquid because the particles in the two liquids have different escaping tendencies. So a *vapour composition* curve can be added to the vapour pressure curves.

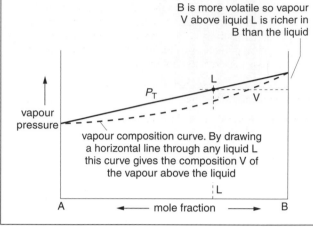

B is more volatile so vapour V above liquid L is richer in B than the liquid

vapour composition curve. By drawing a horizontal line through any liquid L this curve gives the composition V of the vapour above the liquid

Boiling point curves

Because temperature is easier to measure than pressure — let alone partial pressure — it is more useful to have a boiling point curve than a vapour pressure curve. Because a liquid with a low vapour pressure boils at a high temperature and vice versa, the boiling point curve looks like a reflection of the vapour pressure curve.

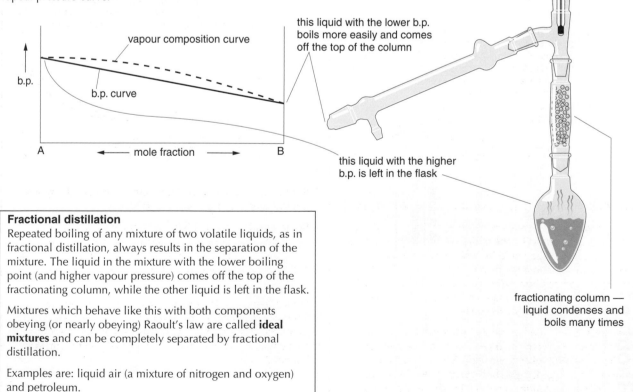

this liquid with the lower b.p. boils more easily and comes off the top of the column

this liquid with the higher b.p. is left in the flask

fractionating column — liquid condenses and boils many times

Fractional distillation

Repeated boiling of any mixture of two volatile liquids, as in fractional distillation, always results in the separation of the mixture. The liquid in the mixture with the lower boiling point (and higher vapour pressure) comes off the top of the fractionating column, while the other liquid is left in the flask.

Mixtures which behave like this with both components obeying (or nearly obeying) Raoult's law are called **ideal mixtures** and can be completely separated by fractional distillation.

Examples are: liquid air (a mixture of nitrogen and oxygen) and petroleum.

Non-ideal liquid mixtures:
ones which do not obey Raoult's Law

Most mixtures do not behave like the ones on the previous page. The particles of one liquid surround those of the other and change the forces on them so affecting their tendency to escape from the liquid.

NEGATIVE DEVIATIONS FROM RAOULT'S LAW

If the attractive forces between the different particles from the two liquids are *stronger* than the attractive forces in the pure liquids, then the paricles will be held in the liquid more strongly. *Fewer* particles will escape; the *vapour pressure will be lower* than that predicted by Raoult.

This happens for example when the hydrogen halides are mixed with water. The ions formed attract molecules strongly.

The vapour pressure curves look like this:

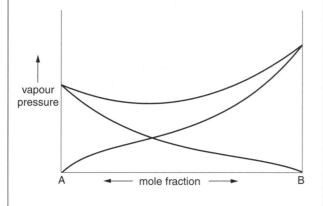

These lead to a set of boiling point and vapour composition curves which look like this:

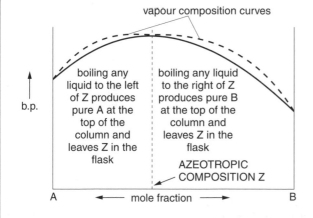

It is helpful to think of this as two ideal curves side by side.

Whichever side of the maximum you start, boiling a mixture like this always results in one of the pure components coming off as vapour until the liquid in the flask reaches the composition of the minimum vapour pressure (called the **azeotropic composition**) when the remainder of the liquid will boil over unchanged. Liquid mixtures which boil like this are called **constant boiling mixtures**.

This means that if you boil dilute hydrochloric acid, pure water comes off and the acid gets more concentrated until it reaches the azeotropic composition. The mixture will then boil over unchanged. If you boil concentrated hydrochloric acid, hydrogen chloride comes off and the acid gets more dilute until it reaches the azeotropic composition.

POSITIVE DEVIATIONS FROM RAOULT'S LAW

If the attractive forces between the different particles from the two liquids are *weaker* than the attractive forces in the pure liquids, then the particles will be held in the liquid less well. *More* particles will escape; the *vapour pressure will be higher* than that predicted by Raoult.

This happens for example when ethanol and water are mixed.

The vapour pressure curves look like this:

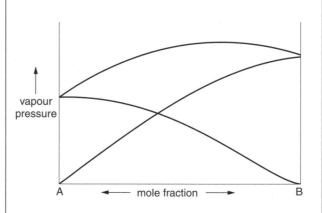

These lead to a set of boiling point and vapour composition curves which look like this:

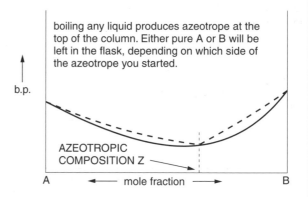

It is helpful to think of this as two ideal curves side by side, but this time they meet at a minimum.

Boiling a mixture like this always results in vapour of the *azeotropic composition* coming off, whatever mixture you started with. Eventually one of the components will run out and the other component will be left pure in the flask.

This means that starting from dilute alcohol — as the result of fermentation — it is never possible to get pure alcohol by fractional distillation. The best that can be achieved is the azeotrope. Pure alcohol can be got from this by adding calcium metal, which reacts with the water in the mixture, but not the ethanol.

Thermodynamics is the study of energy changes during chemical reactions

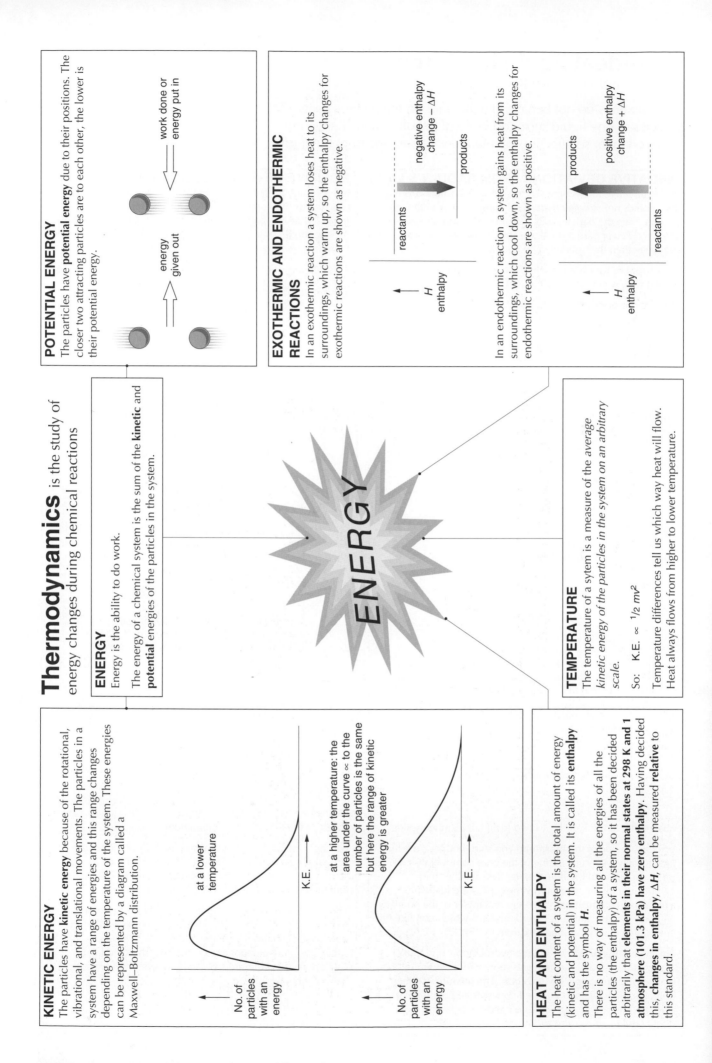

POTENTIAL ENERGY

The particles have **potential energy** due to their positions. The closer two attracting particles are to each other, the lower is their potential energy.

energy given out

work done or energy put in

EXOTHERMIC AND ENDOTHERMIC REACTIONS

In an exothermic reaction a system loses heat to its surroundings, which warm up, so the enthalpy changes for exothermic reactions are shown as negative.

reactants

negative enthalpy change – ΔH

products

In an endothermic reaction a system gains heat from its surroundings, which cool down, so the enthalpy changes for endothermic reactions are shown as positive.

products

positive enthalpy change + ΔH

reactants

H enthalpy

ENERGY

Energy is the ability to do work.

The energy of a chemical system is the sum of the **kinetic** and **potential** energies of the particles in the system.

TEMPERATURE

The temperature of a sytem is a measure of the *average kinetic energy of the particles in the system on an arbitrary scale.*

So: K.E. $\propto \frac{1}{2} mv^2$

Temperature differences tell us which way heat will flow. Heat always flows from higher to lower temperature.

KINETIC ENERGY

The particles have **kinetic energy** because of the rotational, vibrational, and translational movements. The particles in a system have a range of energies and this range changes depending on the temperature of the system. These energies can be represented by a diagram called a Maxwell–Boltzmann distribution.

at a lower temperature

No. of particles with an energy

K.E. ⟶

at a higher temperature: the area under the curve ∝ to the number of particles is the same but here the range of kinetic energy is greater

No. of particles with an energy

K.E. ⟶

HEAT AND ENTHALPY

The heat content of a system is the total amount of energy (kinetic and potential) in the system. It is called its **enthalpy** and has the symbol H.

There is no way of measuring all the energies of all the particles (the enthalpy) of a system, so it has been decided arbitrarily that **elements in their normal states at 298 K and 1 atmosphere (101.3 kPa) have zero enthalpy**. Having decided this, **changes in enthalpy**, ΔH, can be measured **relative to** this standard.

Standard enthalpy changes

So that different enthalpy changes can be compared, other variables are kept constant, so all standard enthalpy changes are measured for a **mole of substance** reacting under the **standard conditions of 289 K and 1 atmosphere (101.3 kPa)**.

Important standard enthalpy changes are given below.

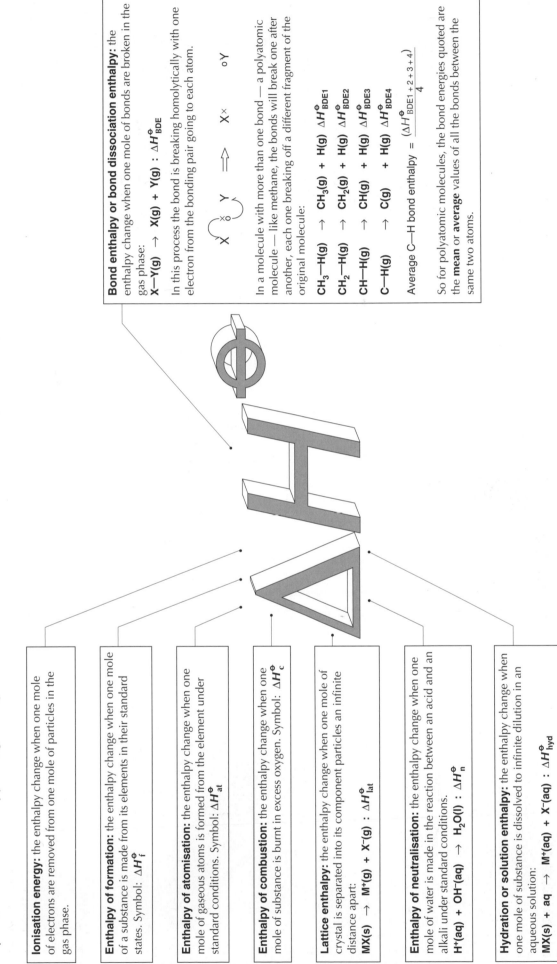

Bond enthalpy or bond dissociation enthalpy: the enthalpy change when one mole of bonds are broken in the gas phase:

$X\!-\!Y(g) \rightarrow X(g) + Y(g) : \Delta H^{\ominus}_{BDE}$

In this process the bond is breaking homolytically with one electron from the bonding pair going to each atom.

In a molecule with more than one bond — a polyatomic molecule — like methane, the bonds will break one after another, each one breaking off a different fragment of the original molecule:

$CH_3\!-\!H(g) \rightarrow CH_3(g) + H(g) \quad \Delta H^{\ominus}_{BDE1}$

$CH_2\!-\!H(g) \rightarrow CH_2(g) + H(g) \quad \Delta H^{\ominus}_{BDE2}$

$CH\!-\!H(g) \rightarrow CH(g) + H(g) \quad \Delta H^{\ominus}_{BDE3}$

$C\!-\!H(g) \rightarrow C(g) + H(g) \quad \Delta H^{\ominus}_{BDE4}$

Average C—H bond enthalpy $= \dfrac{(\Delta H^{\ominus}_{BDE1+2+3+4})}{4}$

So for polyatomic molecules, the bond energies quoted are the **mean** or **average** values of all the bonds between the same two atoms.

Ionisation energy: the enthalpy change when one mole of electrons are removed from one mole of particles in the gas phase.

Enthalpy of formation: the enthalpy change when one mole of a substance is made from its elements in their standard states. Symbol: $\Delta H^{\ominus}_f$

Enthalpy of atomisation: the enthalpy change when one mole of gaseous atoms is formed from the element under standard conditions. Symbol: $\Delta H^{\ominus}_{at}$

Enthalpy of combustion: the enthalpy change when one mole of substance is burnt in excess oxygen. Symbol: $\Delta H^{\ominus}_c$

Lattice enthalpy: the enthalpy change when one mole of crystal is separated into its component particles an infinite distance apart:

$MX(s) \rightarrow M^+(g) + X^-(g) : \Delta H^{\ominus}_{lat}$

Enthalpy of neutralisation: the enthalpy change when one mole of water is made in the reaction between an acid and an alkali under standard conditions.

$H^+(aq) + OH^-(aq) \rightarrow H_2O(l) : \Delta H^{\ominus}_n$

Hydration or solution enthalpy: the enthalpy change when one mole of substance is dissolved to infinite dilution in an aqueous solution:

$MX(s) + aq \rightarrow M^+(aq) + X^-(aq) : \Delta H^{\ominus}_{hyd}$

First law of thermodynamics

or the **law of conservation of energy** states that energy can neither be created nor destroyed, but only transferred from one form to another. It is applied to calorimetry, Hess's law, and energy cycles.

HESS'S LAW

This states that the enthalpy change in a chemical reaction depends only on the initial and final states and is independent of the reaction pathway.

Energy changes are often expressed in the form of a triangle, the sides of which represent the different reaction pathways.

These triangles are used to find enthalpy changes that cannot be measured directly in the laboratory.

For example, they can be based on enthalpies of formation:

$$NH_3 + HCl \xrightarrow{\Delta H_r^\ominus} NH_4Cl$$

with $\Delta H_f^\ominus(HCl)$, $\Delta H_f^\ominus(NH_3)$, $\Delta H_f^\ominus(NH_4Cl)$ and N_2, H_2, Cl_2

$$\Delta H_f^\ominus(NH_3) + \Delta H_f^\ominus(HCl) + \Delta H_r^\ominus = \Delta H_f^\ominus(NH_4Cl)$$

reactants $\xrightarrow{\Delta H^\ominus_{reaction}}$ products

$\Delta H_f^\ominus$(reactants), indirect pathway, $\Delta H_f^\ominus$(products), direct pathway, elements

They can also be based on enthalpies of combustion:

$$CH_3CH_2OH \xrightarrow{\Delta H_r^\ominus} CH_3CHO$$

with $\Delta H_c^\ominus(CH_3CH_2OH)$, $\Delta H_c^\ominus(CH_3CHO)$ and CO_2, H_2O

$$\Delta H_c^\ominus(CH_3CH_2OH) = \Delta H_r^\ominus + \Delta H_c^\ominus(CH_3CHO)$$

reactants $\xrightarrow{\Delta H^\ominus_{reaction}}$ products

$\Delta H_c^\ominus$(reactants), indirect pathway, $\Delta H_c^\ominus$(products), direct pathway, combustion products

BORN–HABER AND OTHER ENERGY CYCLES

These are graphical representations of Hess's law and are normally drawn relative to a **datum line**, which is the arbitrary zero enthalpy content of any pure element. Born–Haber cycles are specifically concerned with the formation of an ionic substance.

Positive enthalpy changes are shown as going up the diagram while negative ones are shown as going down.

REACTANTS — positive enthalpy charge → INTERMEDIATES — negative enthalpy charge → PRODUCTS

Energy ←

CALORIMETRY

When a reaction is carried out in a calorimeter

the heat lost/gained by the reacting system = the heat gained/lost by the calorimeter and its contents

Changes in heat content are calculated using:

ΔH = mass × specific heat capacity × ΔT, the change in temperature

Enthalpies of combustion are measured in a bomb calorimeter, in which known masses are burnt in excess oxygen.

Using energy cycles I

1. TO EXPLAIN THE DIFFERENT STRENGTHS OF ACIDS

Compare the strength of the bond joining the hydrogen to the rest of the acid and then the hydration enthalpies of the ions formed. These two steps give an indication of some of the factors influencing acid strength.

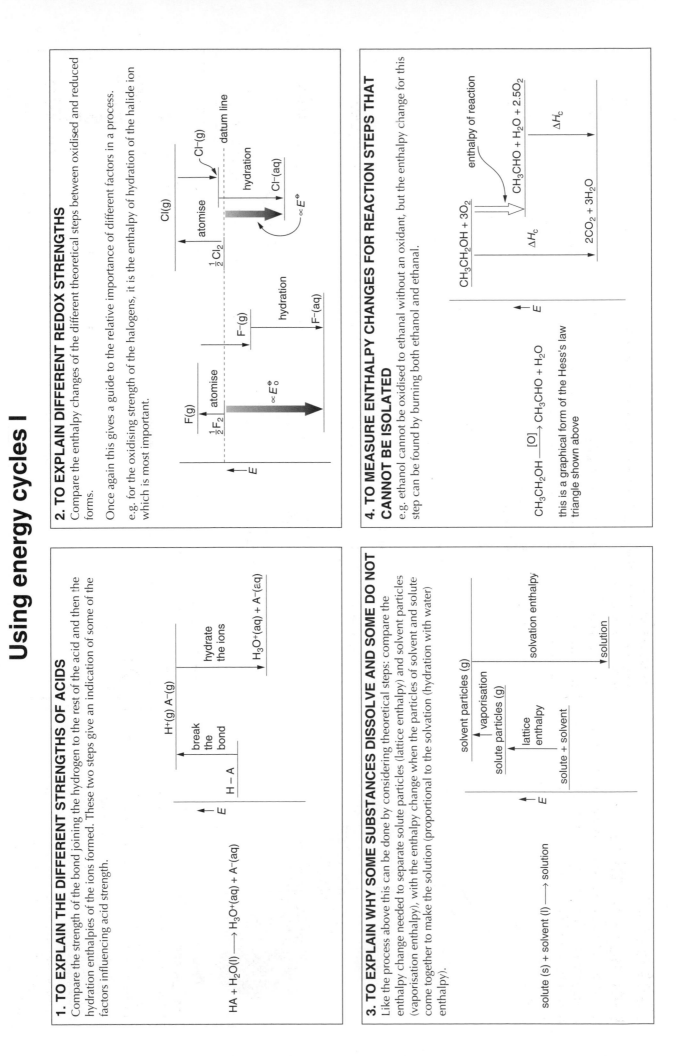

2. TO EXPLAIN DIFFERENT REDOX STRENGTHS

Compare the enthalpy changes of the different theoretical steps between oxidised and reduced forms.

Once again this gives a guide to the relative importance of different factors in a process.

e.g. for the oxidising strength of the halogens, it is the enthalpy of hydration of the halide ion which is most important.

3. TO EXPLAIN WHY SOME SUBSTANCES DISSOLVE AND SOME DO NOT

Like the process above this can be done by considering theoretical steps: compare the enthalpy change needed to separate solute particles (lattice enthalpy) and solvent particles (vaporisation enthalpy), with the enthalpy change when the particles of solvent and solute come together to make the solution (proportional to the solvation (hydration with water) enthalpy).

4. TO MEASURE ENTHALPY CHANGES FOR REACTION STEPS THAT CANNOT BE ISOLATED

e.g. ethanol cannot be oxidised to ethanal without an oxidant, but the enthalpy change for this step can be found by burning both ethanol and ethanal.

$$CH_3CH_2OH \xrightarrow{[O]} CH_3CHO + H_2O$$

this is a graphical form of the Hess's law triangle shown above

Using energy cycles II

5. TO EXPLAIN WHY A SUBSTANCE HAS A PARTICULAR FORMULA

e.g. why is the formula of magnesium chloride $MgCl_2$ and not $MgCl$ or $MgCl_3$? Born–Haber cycles for the three reactions reveal that $MgCl_2$ is thermodynamically more stable than the other two.

6. TO TEST THE BONDING MODEL FOR COVALENT SUBSTANCES LIKE BENZENE

The cyclohexatriene structure does not follow the pattern of cyclohexene and cyclohexadiene, so the real structure of benzene has a delocalised π system.

7. TO CALCULATE IONIC LATTICE ENERGIES

This Born–Haber cycle is constructed by

• making the cations from the solid metal in the correct quantities for the formula of the ionic solid

• then making the anions (remember that although these are ions they involve electron affinities not ionisation energies)

• then bringing the ions together to form a lattice.

This stepwise reaction pathway has to be equal to the direct route of the enthalpy of formation of the solid from its elements, so the lattice energy can be found.

direct pathway = indirect pathway in enthalpy changes

direct formation = formation of ions + lattice energy

8. TO TEST IONIC AND COVALENT BONDING MODELS

Silver chloride made from a metal and a non-metal is predicted to be ionic, but the actual lattice energy (LE), found using a Born–Haber cycle, does not agree with the theoretical one predicted from physical calculations.

This suggests that the ionic model is not the right one for silver chloride. Silver is a d block metal. Electrons in d orbitals shield the nucleus less well than electrons in s or p orbitals. This means that the nucleus of the silver ion is poorly shielded and so has a greater polarising power than expected. The silver ion polarises the chloride ion leading to electron density between the two ions. Silver chloride has a high degree of covalency.

COMPOUND	THEORETICAL LE from physics calculations	ACTUAL LE from Born–Haber cycle
NaCl	−770	−780
NaBr	−735	−742
KCl	−702	−711

agreement good ∴ ionic model a good one here

AgCl	−833	−905
AgBr	−816	−891

agreement poor ∴ ionic model less suitable due to covalent character

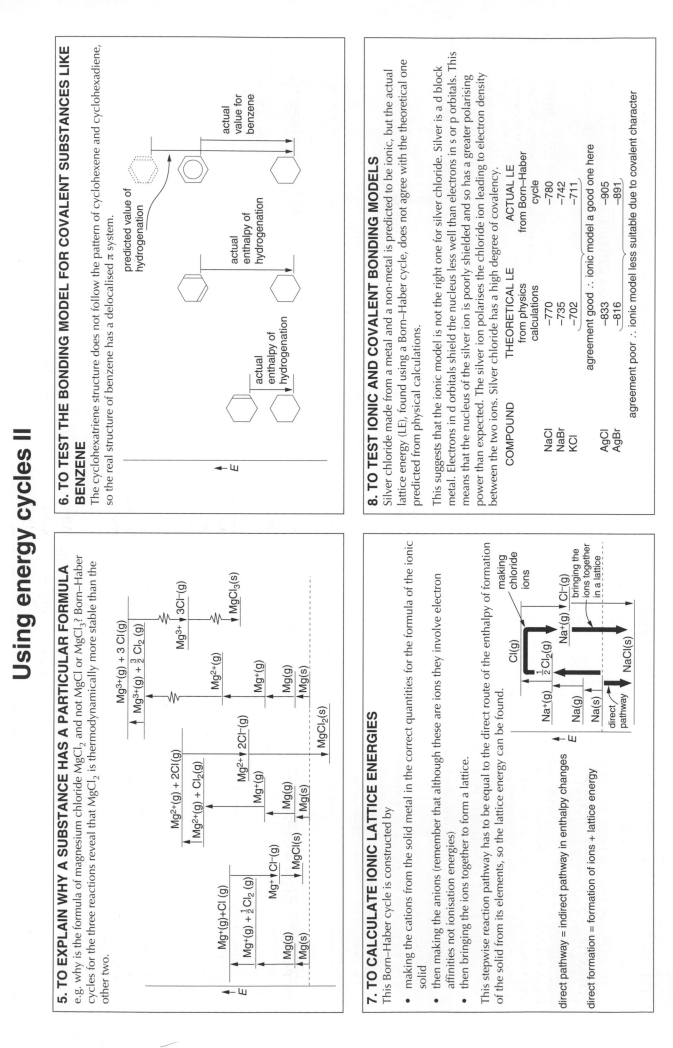

Entropy and Gibbs free energy:
the disorder of molecules and energy available to do work

POTENTIAL AND KINETIC ENERGY

In the arctic almost everything is in a lattice in the solid phase. Particles have very little kinetic energy and so pack closely together in ordered arrangements. Under these conditions of low temperature, as attractive forces between the particles pull them together, the potential energy of the system decreases.

On the sun, everything is in the gas phase. Particles have so much kinetic energy that they overcome all the attractive forces between them and move randomly. So at high temperatures, there is no order and the kinetic energy of the system increases.

Between these two extremes, the ordering tendency of potential energy and the disrupting effect of kinetic energy are in competition. It is helpful to be able to measure these two tendencies and use them to predict whether any particular change will happen.

ENTROPY: A MEASURE OF DISORDER

Exothermic changes, $-\Delta H$, in which a system decreases its potential energy are likely to happen, but not all changes that happen are exothermic.

Why do endothermic changes like evaporation happen? They happen because some of the particles have enough kinetic energy to overcome the forces holding them into the liquid and so they escape. We can also say that they happen because statistically they are more likely to happen. There is only one ordered arrangement, one perfect lattice, for a given system and an almost infinite number of disordered arrangements. Systems just naturally become more disordered or random unless work is done on them to order them — think of your bedroom! — because the disordered state is so much more likely.

The disorder in a system is called its **entropy, S**. A crystal lattice at 0 K has perfect order and therefore zero entropy. The hotter it gets the more kinetic energy the particles in it have and the more options they have for arranging themselves. So the entropy of a system is temperature dependent and increases with temperature.

Finding entropy changes

For a reaction, the total entropy change is given by:

$$\Delta S^{\ominus}_{total} = \Delta S^{\ominus}_{system} + \Delta S^{\ominus}_{surroundings}$$

where $\Delta S^{\ominus}_{system} = \Delta S^{\ominus}_{products} - \Delta S^{\ominus}_{reactants}$ — in $J mol^{-1}$

and $\Delta S^{\ominus}_{surroundings} = -\dfrac{\Delta H^{\ominus}}{T}$ — in Kelvin

PREDICTING WHETHER A REACTION WILL HAPPEN

This equation can be rearranged like this so that we can predict whether a reaction is likely to happen

$$\Delta G = \Delta H - T\Delta S$$

Negative in exothermic changes

Statistically likely to increase, particularly at high temperatures

- Exothermic changes in which ΔH is negative are likely to happen.
- Changes in which the entropy increases, e.g. where small, gaseous molecules are made, are likely to happen, especially at high temperatures where there is lots of disruptive kinetic energy.
- Endothermic changes producing small molecules — like evaporation — may happen and are more likely to at high temperatures.
- Endothermic changes producing large, complex molecules — like photosynthesis — only happen if there is an external energy source — the sun — to drive them.

This can be summarised by saying that reactions in which the change in free energy is negative are likely to happen spontaneously, those in which the change in Gibbs free energy is positive are unlikely to happen unless external work is done on the system.

FREE ENERGY

When a system transfers energy to the surroundings, some of this energy is available to do work. This is called the **free energy** (sometimes the Gibbs free energy). Some of it is involved in rearranging the system, is not available to do work, and is 'unfree energy'

$$\Delta H = \Delta G + T\Delta S$$

Total energy transferred between system and surroundings

Free energy available to do work

'Unfree' energy not available to do work

Kinetics: the facts

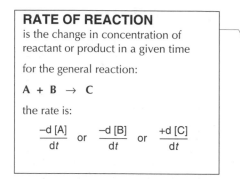

RATE OF REACTION
is the change in concentration of reactant or product in a given time

for the general reaction:

A + B → C

the rate is:

$$\frac{-d\,[A]}{dt} \quad \text{or} \quad \frac{-d\,[B]}{dt} \quad \text{or} \quad \frac{+d\,[C]}{dt}$$

RATE LAW OR KINETICS OF A REACTION
is the equation relating the rate of reaction at any time to the concentration of reactants at that time

e.g. rate is proportional to $[A] \times [B]^2$ or rate $\propto [A][B]^2$

so rate $= k[A][B]^2$ where k is the constant of proportionality or **rate constant** for this reaction

This leads to the concept of **order**.

ORDER
is the number of concentration factors in the rate equation.

In the example above the order with respect to A is 1 and with respect to B is 2; the overall order is 1 + 2 = 3.

The order of a reaction can only be found by experiment and cannot be worked out from the equation of the reaction.

Common orders
zero order: rate is unchanged with time: rate $\propto [A]^0$

first order: rate is directly proportional to one concentration term: rate $\propto [A]^1$

second order: rate is proportional to two concentration terms:
rate $\propto [A]^1[B]^1$ or rate $\propto [A]^2$

EXPERIMENTS TO FIND ORDER:

discontinuous	continuous
many separate experiments with different starting concentrations	one experiment
one reading per experiment	many readings as experiment goes on
e.g. clock reactions; thiosulphate and acid	e.g. gas syringes; sampling experiments

FACTORS THAT AFFECT THE RATE OF REACTION
Reaction rate is affected by:
- the concentration of the reactants (and pressure in gas phase reactions)
- the particle size in heterogeneous reactions (those involving solids with gases or liquids)
- the temperature of the reacting system — typically the rate doubles for every 10 °C rise in temperature. (Some reactions are affected by light energy instead of heat.)
- the addition of a suitable catalyst

RATE DETERMINING STEPS
In a multistep reaction, the slowest step controls the rate.

CHAIN REACTIONS
are reactions in which each step produces the reactant for the next step

DETERMINING ORDERS AND RATE CONSTANTS
For **discontinuous experiments**: inspect the data to see how changing concentration affects the rate (see example on p. 56).

Once order is found, write a rate equation then substitute one set of concentrations in to find the rate constant.

For **continuous experiments**

either: 1. Plot the reactant concentration against time.

2. Is it a straight line? If so then the order is zero.

3. If not order is first or second, measure and tabulate half lives.

4. Are they constant? If so, then the order is first and rate constant is $k = \dfrac{\log_e 2}{\text{half life}}$

5. If not, work out the initial concentration c_0 times the half life for several values of half life.

6. Are they constant? If so then the order is second.

or: 1. Plot the reactant concentration against time.

2. Is it a straight line? If so then the order is zero.

3. If not, plot ln[reactant] against time.

4. Is it a straight line? If so the order is first and the gradient = the rate constant k.

5. If not, plot the reciprocal of [reactant] against time.

6. Is it a straight line? If so the order is second and the gradient = the rate constant k.

Kinetics: the theory

THE COLLISION THEORY

This states that to react

- particles must collide
- with enough energy to break existing bonds
- and with the correct orientation to bring reactive sites close together

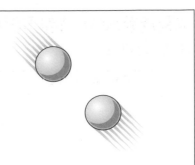

RELATING THE THEORY TO THE FACTORS AFFECTING RATE

Changes in concentration (or pressure for a gas) change the number of particles in a unit volume and hence the number of collisions per unit time in that volume. If the number of collisions changes the rate will change.

Surface area changes in heterogeneous reactions change the **number of collisions** between the fluid phase (liquid or gas) and the solid surface. Once again, if the number of collisions changes then the rate will also change.

Changes in temperature change the kinetic energy of the particles and hence the number of successful collisions with enough energy to break existing bonds and make product particles. The minimum energy needed for a successful collision is called the activation energy.

Increasing the temperature of the system:

1. increases the range of kinetic energies;
2. increases the average kinetic energy;
3. increases the population of particles with more than the activation energy (shown by the shaded areas under the graph).

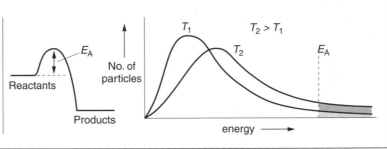

Addition of a catalyst can decrease the required activation energy so that a greater population of particles will collide successfully.

A catalyst increases the rate without being used up. It does this by providing a reaction pathway with a lower activation energy through:

1. making more collisions have favourable orientation;
2. locally increasing concentrations on its surfaces;
3. providing a series of simple steps rather than one high energy one;
4. providing a better attacking group which is regenerated later;
5. increasing the reactivity of the reactive site.

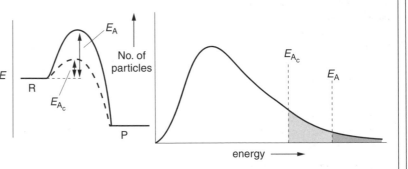

Relating the theory to the rate equation

$$\text{Rate} = k\,[A]^x[B]^y$$

concentration changes change rate

$$k = Ae^{-E_A/RT}$$

k is temperature dependent

and k depends on the activation energy and so is affected by adding a catalyst

This equation is often referred to as the **Arrhenius equation**. A, the Arrhenius constant, is a 'collision factor'.

Rate determining steps

Like roadworks on a motorway, the slowest step in a multistep reaction controls the overall rate.

e.g.

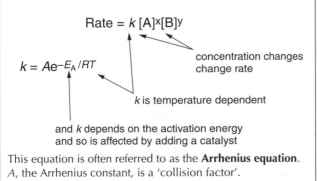

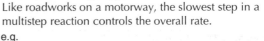

so rate depends on [propanone] not [iodine]
reaction is 1st order w.r.t. propanone, zero w.r.t. iodine

Experimental rate methods

DISCONTINUOUS METHODS
e.g. Clock reactions
(remember that time is inversely proportional to rate)

view through solution

time

cross on paper viewed through solution

clock

dilute acid and sodium thiosulphate solution

Analysing the data

[A] mol dm⁻³	[B] mol dm⁻³	Rate mol dm⁻³ s⁻¹
0.1	0.1	1.2×10^{-3}
0.2	0.1	2.4×10^{-3}
0.2	0.3	2.4×10^{-3}

doubling [A] doubles rate ∴ reaction is FIRST order w.r.t. [A]

changing [B] does not change rate ∴ reaction is ZERO order w.r.t. [B]

so the rate equation is: Rate $\propto [A]^1 [B]^0$
Rate $= k [A]^1 [B]^0$

To find the value of k substitute in one set of data:
$1.2 \times 10^{-3} = k \ 0.1 \times 1$
$k = 1.2 \times 10^{-2}$

To find units of k: the units on each side of the equation must balance — substitute in the units

mol dm⁻³ s⁻¹$= k$ mol dm⁻³
k must have units of s⁻¹

CONTINUOUS METHODS

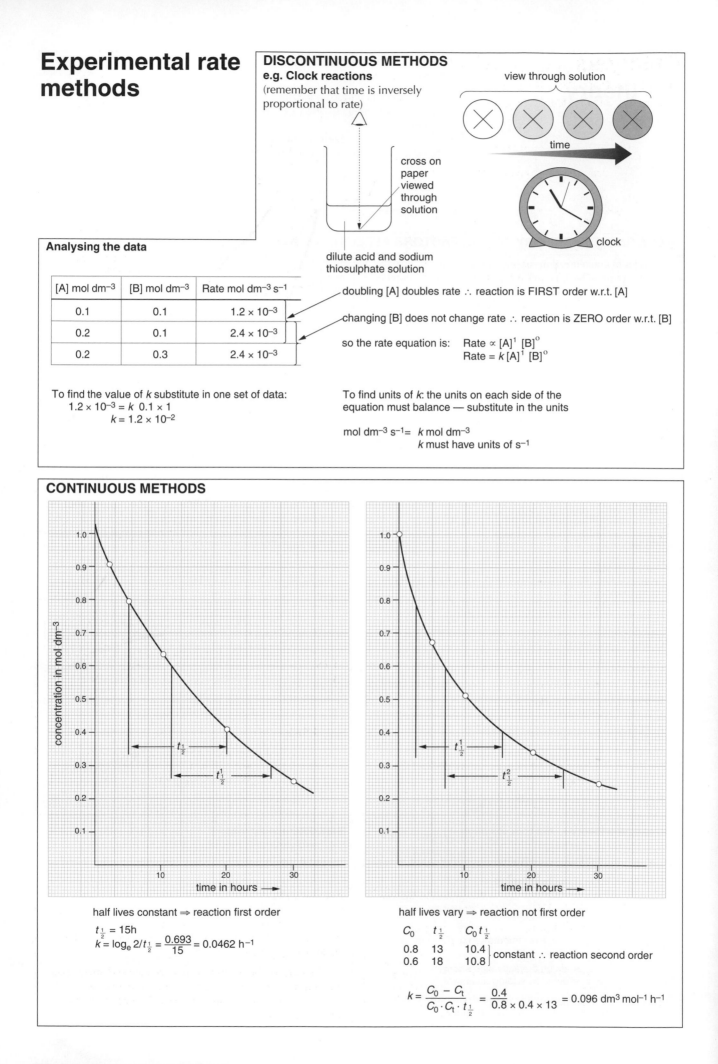

half lives constant ⇒ reaction first order

$t_{\frac{1}{2}} = 15h$
$k = \log_e 2/t_{\frac{1}{2}} = \dfrac{0.693}{15} = 0.0462 \ h^{-1}$

half lives vary ⇒ reaction not first order

C_0	$t_{\frac{1}{2}}$	$C_0 t_{\frac{1}{2}}$
0.8	13	10.4
0.6	18	10.8

constant ∴ reaction second order

$k = \dfrac{C_0 - C_t}{C_0 \cdot C_t \cdot t_{\frac{1}{2}}} = \dfrac{0.4}{0.8 \times 0.4 \times 13} = 0.096 \ dm^3 \ mol^{-1} \ h^{-1}$

Catalysis: changing the rate of reactions by adding a substance which does not get used up

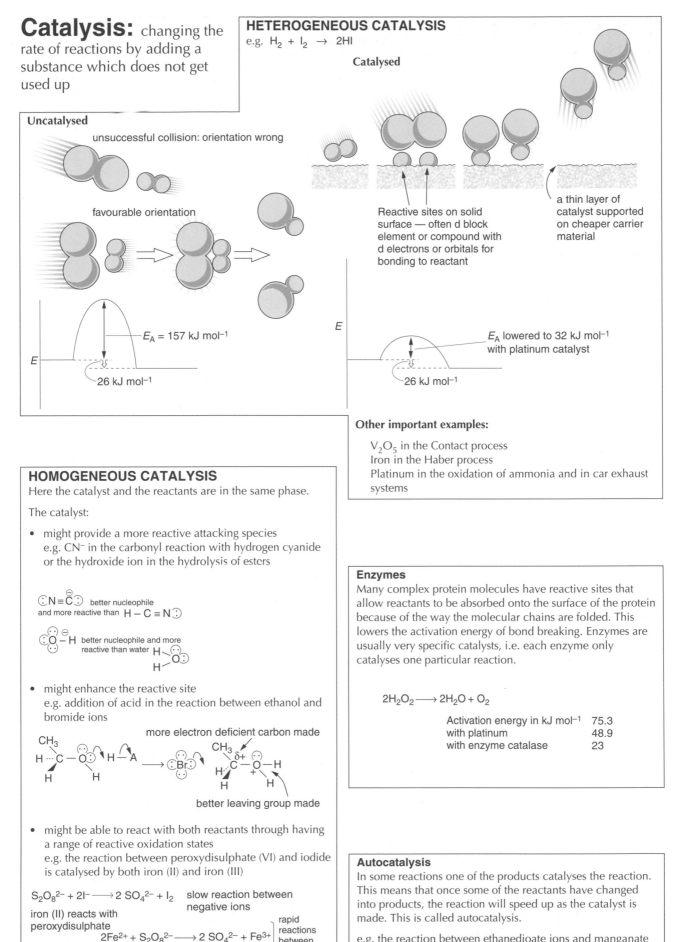

HETEROGENEOUS CATALYSIS
e.g. $H_2 + I_2 \rightarrow 2HI$

Catalysed

Reactive sites on solid surface — often d block element or compound with d electrons or orbitals for bonding to reactant

a thin layer of catalyst supported on cheaper carrier material

E_A lowered to 32 kJ mol^{-1} with platinum catalyst

26 kJ mol^{-1}

Uncatalysed

unsuccessful collision: orientation wrong

favourable orientation

$E_A = 157$ kJ mol^{-1}

26 kJ mol^{-1}

Other important examples:

V_2O_5 in the Contact process
Iron in the Haber process
Platinum in the oxidation of ammonia and in car exhaust systems

HOMOGENEOUS CATALYSIS
Here the catalyst and the reactants are in the same phase.

The catalyst:

- might provide a more reactive attacking species
  e.g. CN^- in the carbonyl reaction with hydrogen cyanide or the hydroxide ion in the hydrolysis of esters

$N \equiv C$ better nucleophile and more reactive than $H - C \equiv N$

$O - H$ better nucleophile and more reactive than water $H - O - H$

- might enhance the reactive site
  e.g. addition of acid in the reaction between ethanol and bromide ions

more electron deficient carbon made

better leaving group made

- might be able to react with both reactants through having a range of reactive oxidation states
  e.g. the reaction between peroxydisulphate (VI) and iodide is catalysed by both iron (II) and iron (III)

$S_2O_8^{2-} + 2I^- \longrightarrow 2\,SO_4^{2-} + I_2$ slow reaction between negative ions

iron (II) reacts with peroxydisulphate
$2Fe^{2+} + S_2O_8^{2-} \longrightarrow 2\,SO_4^{2-} + Fe^{3+}$

iron (III) reacts with iodide ions $2Fe^{3+} + 2I^- \longrightarrow 2\,Fe^{2+} + I_2$

rapid reactions between oppositely charged ions

Enzymes
Many complex protein molecules have reactive sites that allow reactants to be absorbed onto the surface of the protein because of the way the molecular chains are folded. This lowers the activation energy of bond breaking. Enzymes are usually very specific catalysts, i.e. each enzyme only catalyses one particular reaction.

$2H_2O_2 \longrightarrow 2H_2O + O_2$

Activation energy in kJ mol^{-1}	75.3
with platinum	48.9
with enzyme catalase	23

Autocatalysis
In some reactions one of the products catalyses the reaction. This means that once some of the reactants have changed into products, the reaction will speed up as the catalyst is made. This is called autocatalysis.

e.g. the reaction between ethanedioate ions and manganate (VII) ions results in the manganese being reduced to manganese (II) ions. These manganese (II) ions catalyse the original reaction.

Chemical equilibrium:
the study of reversible reactions

REVERSIBLE REACTIONS

Reversible reactions make products which themselves react to give back the products. These reactions never stop because once some product is made it can regenerate the reactants from which it came.

e.g. acid reacts with alcohol to make ester and water, but ester reacts with water to make acid and alcohol

To show a system like this the equilibrium sign $\rightleftharpoons$ is used

acid + alcohol $\rightleftharpoons$ ester + water

Look at the way concentration and rate change with time here.

equilibrium reached: no further change

concentration of product

concentration of reactant

conc

time

equilibrium reached

forward rate

back rate

rate

time

After time t, products are being made at exactly the same rate as they are reacting to make reactants. Because the two opposite rates are exactly equal there is no external change and the system is said to be in **dynamic equilibrium**.

REACTIONS WHICH GO TO COMPLETION

Many reactions continue until one of the reactants runs out and then the reaction stops.

e.g. a fire burns until the fuel runs out
marble chips react with acid until either the acid or the marble is used up

These are examples of reactions which go to completion as the graph of concentration against time shows.

reaction stops when concentration reaches zero

conc

time

RECOGNISING EQUILIBRIUM SYSTEMS

1. They can be approached from either end.

e.g. chromate ions can be changed into dichromate ions by adding acid

and dichromate ions can be turned into chromate ions by adding hydroxide ions, which remove the hydroxonium ions.

$$2CrO_4^{2-} + 2\,H_3O^+ \xrightarrow[\text{add base}]{\text{add acid}} Cr_2O_7^{2-} + 3H_2O$$

2. If the temperature is changed an equilibrium system changes, but if the original temperature is restored, the system will go back to its original state.

PROVING THE DYNAMIC NATURE OF AN EQUILIBRIUM SYSTEM

This is done using radioactive tracers:

1. Set up an equilibrium system and allow it to reach equilibrium

2. By sampling and measuring, find the exact composition of the whole system

3. Now set up an identical system by adding the exact amount you have just measured, *but with one component made of a radioactive isotope*, say one of the reactants

4. Leave the system for a time and then sample again and show that there are radioactive products, thus proving that even at equilibrium matter is being converted either way

The equilibrium law:

relates the concentrations of reactants and products

THE EQUILIBRIUM LAW

The equilibrium law states that for any system in equilibrium, there is a numerical relationship between the concentrations of the products, raised to the power of their stoichiometric numbers, and the concentrations of the reactants, raised to the powers of their stoichiometric numbers. This relationship is called the **equilibrium constant, K_c** (when trying to explain this in an answer you must give an example like this)

e.g. for the system

$$N_2 + 3H_2 \rightleftharpoons 2NH_3$$

the equilibrium law states that:

$$K_c = \frac{[NH_3]^2}{[N_2] \times [H_2]^3}$$

square brackets, [ ], mean concentration of whatever is inside them.

THE EQUILIBRIUM CONSTANT, K_c, FOR HOMOGENEOUS LIQUID SYSTEMS

Many equilibrium systems are made up of ions in solution. They are all in the same liquid phase. Here K_c is written in terms of concentration. For example:

$$Fe^{3+}(aq) + SCN^-(aq) \rightleftharpoons FeSCN^{2-}(aq)$$

$$K_c = \frac{[FeSCN^{2-}(aq)]}{[Fe^{3+}(aq)][SCN^-(aq)]}$$

$$Cu^{2+}(aq) + 4Cl^-(aq) \rightleftharpoons CuCl_4^{2-}(aq)$$

$$K_c = \frac{[CuCl_4^{2-}(aq)]}{[Cu^{2+}(aq)][Cl^-(aq)]^4}$$

PRODUCTS

REACTANTS

THE EQUILIBRIUM CONSTANT, K_p, FOR HOMOGENEOUS GASEOUS SYSTEMS

For gases it is usually more convenient to measure the pressure of the gas than its concentration. In a mixture of gases the gas is causing only part of the pressure, so the idea of **partial pressure** is used.

The partial pressure of a gas in a mixture is the pressure that the gas would exert if it alone occupied the space containing the mixture.

The pressure caused by a gas is proportional to the number of particle of the gas so we can write:

$$\frac{\text{partial pressure of}}{\text{total pressure of the}} = \frac{\text{number of particles of}}{\text{total number of particles in}}$$
$$\frac{\text{the gas, } p_g}{\text{gas mixture, } P_T} = \frac{\text{the gas, } n_g}{\text{the mixture, } N_T}$$

which can be rearranged to give $p_g = \frac{n_g}{N_T} \times P_T$

n_g/N_T is the mole fraction of the gas

So the partial pressure of a gas equals the mole fraction of that gas multiplied by the total pressure.

Using partial pressures, a form of the equilibrium law can be written in terms of a different equilibrium constant K_p. For example:

$$N_2 + 3H_2 \rightleftharpoons 2NH_3 \qquad 2SO_2 + O_2 \rightleftharpoons 2SO_3$$

$$K_p = \frac{p_{NH_3}^2}{p_{N_2} \times p_{H_2}^3} \qquad K_p = \frac{p_{SO_3}^2}{p_{SO_2}^2 \times p_{O_2}}$$

THE EQUILIBRIUM CONSTANT FOR HETEROGENEOUS SYSTEMS

Many systems contain more than one phase and so are heterogeneous. If one of the phases is a pure solid or liquid, then although the **amount** of the solid or liquid may change, its **concentration** will not. In these cases it is usual to write an equilibrium law expression that does not contain the pure solid or liquid phase's concentration (which is actually included in the modified equilibrium constant). These examples show this point:

$$CaCO_3(s) \rightleftharpoons CaO(s) + CO_2(g)$$

$$K_c = [CO_2] \text{ and } K_p = p_{CO_2}$$

$$Fe_2O_3(s) + 3CO(g) \rightleftharpoons 2Fe(s) + 3CO_2(g)$$

$$K_c = \frac{[CO_2]^3}{[CO]^3} \text{ and } K_p = \frac{p_{CO_2}^3}{p_{CO}^3}$$

The equilibrium constant

THE SIZE OF THE EQUILIBRIUM CONSTANT

The value or size of K_c, tells us whether there are more reactants or products in the system at equilibrium. So

if K_c is $> 10^2$, there will be far more products than reactants
if K_c is $< 10^{-2}$, there will be far more reactants than products
if K_c is between 10^{-2} and 10^2, both reactants and products will be in the system in noticeable amounts.

THE UNITS OF THE EQUILIBRIUM CONSTANT

Equilibrium constants for different systems can have different units. Just remember that the units of concentration are mol dm^{-3} and that the units on each side of the equation must balance, so:

for the system: $N_2 + 3H_2 = 2NH_3$: $K_c = \dfrac{[NH_3]^2}{[N_2] \times [H_2]^3}$

so the units of K_c will be $= \dfrac{\text{mol dm}^{-3} \times \text{mol dm}^{-3}}{\text{mol dm}^{-3} \times \text{mol dm}^{-3} \times \text{mol dm}^{-3} \times \text{mol dm}^{-3}} = \text{dm}^6\ \text{mol}^{-2}$

CALCULATIONS INVOLVING THE EQUILIBRIUM CONSTANT

The equilibrium law is a mathematical expression relating the concentrations of reactants and products. Many questions are set using the relationship (an equation) in which some values are given and one has to be calculated by solving the equation. These often involve solving a quadratic equation.

e.g. The equilibrium constant for the reaction between ethanol and ethanoic acid at room temperature is 4. Calculate the equilibrium amounts of reactants and products when two moles of ethanol and one of ethanoic acid react and reach equilibrium.

$$CH_3CH_2OH + CH_3COOH \rightleftharpoons CH_3COOCH_2CH_3 + H_2O$$

initial amounts	2	1	0	0	moles	Let x moles react
equilibrium amounts	$2-x$	$1-x$	x	x	moles	If this is in a volume V
equilibrium conc.	$\dfrac{2-x}{V}$	$\dfrac{1-x}{V}$	$\dfrac{x}{V}$	$\dfrac{x}{V}$	moles dm^{-3}	

apply the equilibrium law:

$$K_c = \frac{[CH_3COOCH_2CH_3][H_2O]}{[CH_3CH_2OH][CH_3COOH]} = \frac{x \times x}{(2-x)(1-x)} = 4$$

$\Rightarrow x^2 = 4(2-x)(1-x)$

$\Rightarrow x^2 = 8 - 12x + 4x^2$

$\Rightarrow 3x^2 - 12x + 8 = 0$ this is a quadratic like $ax^2 + bx + c = 0$ and can be solved using

$$x = \frac{-b \pm \sqrt{b^2 - 4ac}}{2a}$$

$$= \frac{12 \pm \sqrt{144 - 96}}{6}$$

$$x = \frac{12 \pm \sqrt{48}}{6}$$

$$= \frac{12 \pm 6.92}{6} = 3.15 \text{ or } 0.85$$

the first root is not applicable in this case because x cannot be > 2 (the starting number of moles); so the amounts in the equilibrium mixture are:

ethanol = $2 - x = 1.15$ moles
ethanoic acid = 0.15 moles
ethyl ethanoate = water = 0.85 moles.

THE EQUILIBRIUM POSITION

The **position** of an equilibrium system is a term used to describe qualitatively what the equilibrium constant does quantitatively. It gives an indication of whether the reactants or products are more plentiful in the system. Remember, reactants are written on the left, products on the right in an equation, so:

• if the position lies to the left then the reactants dominate:

$$CH_3COOH(l) + H_2O(l) \rightleftharpoons CH_3COO^-(aq) + H_3O^+(aq) \qquad K_c = 1 \times 10^{-5}$$

• if the position lies to the right the products dominate:

$$HCl(g) + H_2O(l) \rightleftharpoons H_3O^+(aq) + Cl^-(aq) \qquad K_c = 5.55 \times 10^8$$

The effect of changing conditions of equilibrium systems I

Because an equilibrium system is dynamic, with forward and back reactions going on all the time, when the conditions are changed the system usually responds and changes itself. We can work out what happens either by thinking about the equilibrium law expression (better) or by thinking about the relative rates of the forward and back reactions.

Look at the synthesis of ammonia at 600 K as an example:

$$N_2 + 3H_2 \rightleftharpoons 2NH_3 : K_c = 2 \text{ dm}^6 \text{ mol}^{-2}$$

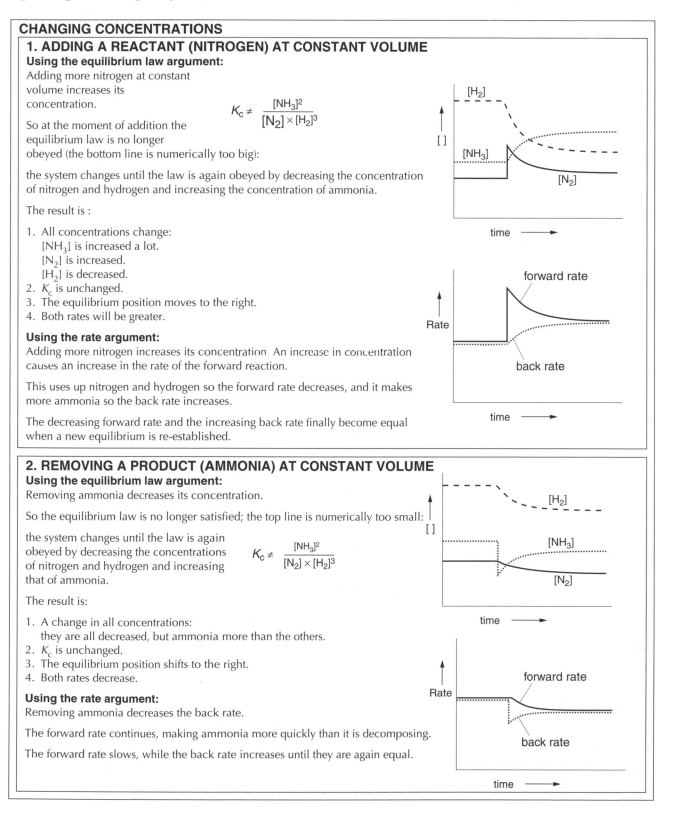

CHANGING CONCENTRATIONS

1. ADDING A REACTANT (NITROGEN) AT CONSTANT VOLUME
Using the equilibrium law argument:
Adding more nitrogen at constant volume increases its concentration.

$$K_c \neq \frac{[NH_3]^2}{[N_2] \times [H_2]^3}$$

So at the moment of addition the equilibrium law is no longer obeyed (the bottom line is numerically too big):

the system changes until the law is again obeyed by decreasing the concentration of nitrogen and hydrogen and increasing the concentration of ammonia.

The result is :

1. All concentrations change:
   $[NH_3]$ is increased a lot.
   $[N_2]$ is increased.
   $[H_2]$ is decreased.
2. K_c is unchanged.
3. The equilibrium position moves to the right.
4. Both rates will be greater.

Using the rate argument:
Adding more nitrogen increases its concentration. An increase in concentration causes an increase in the rate of the forward reaction.

This uses up nitrogen and hydrogen so the forward rate decreases, and it makes more ammonia so the back rate increases.

The decreasing forward rate and the increasing back rate finally become equal when a new equilibrium is re-established.

2. REMOVING A PRODUCT (AMMONIA) AT CONSTANT VOLUME
Using the equilibrium law argument:
Removing ammonia decreases its concentration.

So the equilibrium law is no longer satisfied; the top line is numerically too small:

the system changes until the law is again obeyed by decreasing the concentrations of nitrogen and hydrogen and increasing that of ammonia.

$$K_c \neq \frac{[NH_3]^2}{[N_2] \times [H_2]^3}$$

The result is:

1. A change in all concentrations:
   they are all decreased, but ammonia more than the others.
2. K_c is unchanged.
3. The equilibrium position shifts to the right.
4. Both rates decrease.

Using the rate argument:
Removing ammonia decreases the back rate.

The forward rate continues, making ammonia more quickly than it is decomposing.

The forward rate slows, while the back rate increases until they are again equal.

The effect of changing conditions of equilibrium systems II

3. INCREASING THE PRESSURE ON THE SYSTEM

Using the equilibrium law argument:

Doubling the pressure doubles all the concentrations (assuming the gas is ideal), but these appear in the equilibrium law expression raised to different powers. The top line has two concentration terms while the bottom one has four; so the numerical value of the top line will be less than the bottom line.

$$K_c \neq \frac{[NH_3] \times [NH_3]}{[N_2] \times [H_2] \times [H_2] \times [H_2]}$$

each term has increased by the same amount

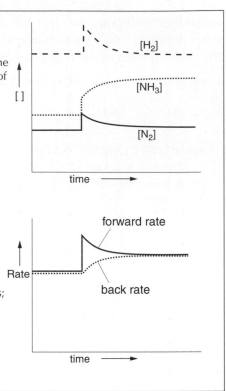

The system reacts, reducing concentrations of reactants and making more products.

The result is:

1. An increase in all concentrations, but more for ammonia than the two reactants.
2. K_c is unchanged.
3. The equilibrium position moves to the right.
4. Both rates will be greater.

Using the rate argument:

The forward reaction depends on collisions between nitrogen and hydrogen molecules; these will happen more often if the gases are compressed into a smaller volume. The back reaction, the decomposition of ammonia molecules, does not need a collision and so will not, at first, be changed by the increase in pressure. As more ammonia gets made, the back rate increases until it equals the forward rate and equilibrium is reached once again.

4. ADDING A NOBLE GAS AT CONSTANT VOLUME

If the volume of the container remains the same, then none of the concentrations will change and so there will be no change in the system.

5. ADDING A NOBLE GAS AT CONSTANT PRESSURE

If the pressure remains the same as the gas is added, then the whole system must expand. This means that all the other gases are spread through a greater volume and so their concentrations are decreased as are their partial pressures. The system will change.

Using the equilibrium law argument

The concentrations of all the gases will fall by the same amount, but they appear in the equilibrium law expression raised to different powers. The top line has two concentration terms in it while the bottom one has four, so the numerical value of the bottom line will decrease more than the top line.

$$K_c \neq \frac{[NH_3] \times [NH_3]}{[N_2] \times [H_2] \times [H_2] \times [H_2]}$$

all decreased by the same amount

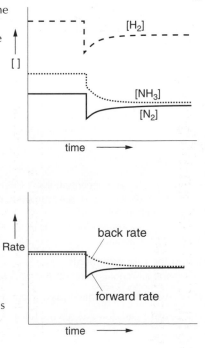

The system reacts, reducing the concentration of the ammonia on the top line and increasing the concentrations of nitrogen and hydrogen on the bottom line until equilibrium is re-established.

The result is:
1. All concentrations decrease, but nitrogen and hydrogen more than ammonia.
2. K_c is unchanged.
3. The equilibrium position moves to the left.
4. Both rates will be slower.

Using the rate argument

The forward rate depends on collisions between nitrogen and hydrogen. Because the molecules are now in a bigger volume they will collide less often so the forward rate decreases. The back rate depends only on the number of ammonia molecules and this has not changed. So for a time the forward rate will be less than the back rate. This means that ammonia is being used up quicker than it is being made. The back rate drops until it once again equals the forward one and a new equilibrium is reached.

The effect of changing conditions of equilibrium systems III

LE CHATELIER'S PRINCIPLE

This is a helpful general rule for checking the changes in equilibrium systems. It states:

If a constraint is applied to a system at equilibrium, the system will react in such a way as to tend to oppose that applied constraint.

So if a product is removed, the position of the system will shift to the right, replacing some of the product. But if a reactant is added, the position of the system moves to the right also, using up some of the added reactant.

If a gaseous system is compressed, the position of the system will move to the side with the fewest moles so reducing the pressure in the system.

If a system is heated it will shift its position towards the endothermic side so absorbing some of the added heat.

Le Chatelier's principle is only ever a general guide and never a reason for a system to change.

6. CHANGING THE TEMPERATURE OF AN EQUILIBRIUM SYSTEM

Qualitative explanation

Changing the temperature of an equilibrium changes the energy content of both reactants and products relative to each other. Both forward and backward rates will change, but not in the same way. The result is that the equilibrium constant and the position of equilibrium will change and so will individual concentrations.

A simple idea to remember is that increasing the temperature pushes the equilibrium position in the endothermic direction. i.e. more energy helps the reaction go uphill.

Quantitative explanation

A more rigorous way to look at temperature changes is based on the relationship between the equilibrium constant, K_c, and the free energy, ΔG, of the reaction.

$$\Delta G = -RT\log_e K, \text{ where } K \approx K_c$$

but we have seen under the energy section that

$$\Delta G = \Delta H - T\Delta S$$

so we can substitute for ΔG in the first equation

$$\Delta H - T\Delta S = -RT\log_e K$$

now if we divide through by RT we get

$$\log_e K = -\Delta H/RT + \Delta S/R,$$

this last term is constant so we get

$$\log_e K = -\Delta H/RT + \text{a constant}$$

Thus there is a direct relationship between the enthalpy change of a reaction and its equilibrium constant. This relationship, which is shown below graphically, can be stated as:

if the forward reaction is exothermic, increasing the temperature will decrease the equilibrium constant and vice versa;

if the forward reaction is endothermic, increasing the temperature will increase the equilibrium constant and vice versa.

Applying these ideas to the ammonia synthesis system:

$$N_2 + 3H_2 \rightleftharpoons 2NH_3 \qquad \Delta H = -92.4 \text{ kJ}$$

An increase in temperature will favour the back reaction, the endothermic one, more than the forward one.

The result is:

1. The concentrations of nitrogen and hydrogen will increase and the concentration of ammonia will decrease.
2. K_c and K_p will decrease.
3. The position of equilibrium will move to the left.
4. Both rates will increase.

EXOTHERMIC FORWARD REACTION

Raising the temperature favours reactants

ENDOTHERMIC FORWARD REACTION

Raising the temperature favours the products

Chromatography and the partition law

The partition (or distribution) law states that when a solute is added to two phases in contact with each other, the ratio of the concentration of the solute in the two phases is constant whatever the amount of solute or each phase.

aqueous layer

trichloroethane layer

$$K = \frac{[I_2]aq}{[I_2]CH_3CCl_3}$$

This applies to solid solutes like iodine added to water/organic mixtures, to oxygen distributed between the air and water, and to solutes distributed between stationary and moving phases in chromatography.

air

water

$$\frac{[O_2]air}{[O_2]aq} = K$$

GLC stands for *Gas Liquid Chromatography* in which small samples of a liquid mixture are injected into a stream of carrier gas passing over some surface active solid material (aluminium oxide, activated charcoal, etc.) packed into a long thin tube. As it passes along the tube the mixture is separated. The end of the tube may lead straight into a mass spectrometer or some other detection device such as a flame photometer.

Fraction collector (optional)

Electronics

Recorder

Detector

Column

Oven

Sample insertion

Flow meter

Carrier gas

TLC stands for *Thin Layer Chromatography* in which small samples (drops) are placed on some absorbent material spread thinly on a carrier surface which may be paper or glass. The paper or glass plate is then placed in a shallow dish of a special solvent which flows over the surface separating the original sample.

Each component in the original sample can then be recognised by measuring the distance it has travelled compared to the distance the solvent has travelled. This is known as the R_f value.

$$R_{f_x} = \frac{d_x}{d_s}$$

d_x = distance travelled by solute x
d_s = distance travelled by solvent

Suspended plate

Glass cylinder

Solvent front

d_x

d_s

Organic solvent with water

Chromatography is the process of separating a solution containing many solutes, which are often present in very small amounts, by placing the solution on some absorbent medium and passing yet another solute across the medium.

Each solute in the original mixture will distribute itself between the solution on the absorbent, stationary medium and the second moving solvent in a unique way. The result is that some solutes will travel across the absorbent medium faster than others, leaving the individual solutes separated and spread.

solute in stationary phase

solvent

GLC, TLC, etc. are initials for different types of chromatography.

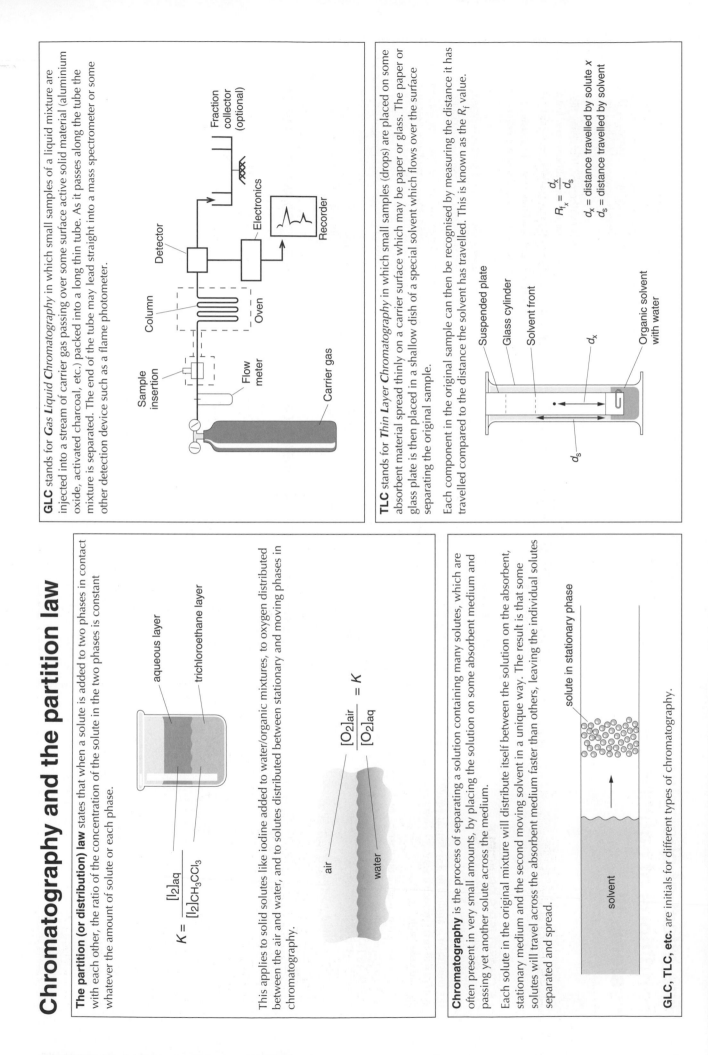

Acids and bases

Key theory: Brønsted–Lowry theory of acids and bases:
Acid–base reactions are ones in which protons are given and taken.
Acids are proton donors
Bases are proton acceptors

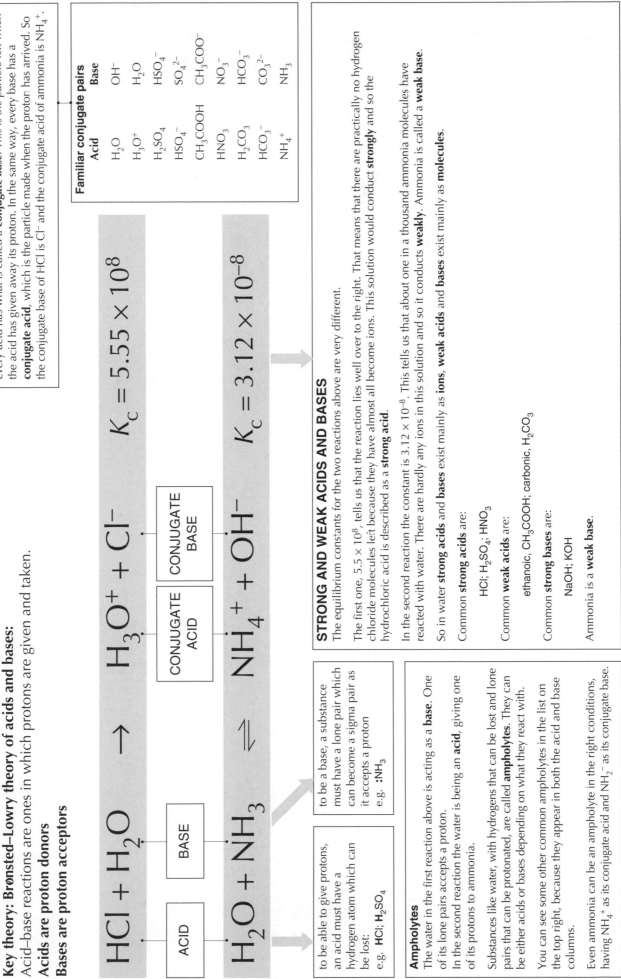

Conjugate pairs

Every acid has what is called a **conjugate base**. This is the particle left when the acid has given away its proton. In the same way, every base has a **conjugate acid**, which is the particle made when the proton has arrived. So the conjugate base of HCl is Cl⁻ and the conjugate acid of ammonia is NH_4^+.

Familiar conjugate pairs

Acid	Base
H_2O	OH^-
H_3O^+	H_2O
H_2SO_4	HSO_4^-
HSO_4^-	SO_4^{2-}
CH_3COOH	CH_3COO^-
HNO_3	NO_3^-
H_2CO_3	HCO_3^-
HCO_3^-	CO_3^{2-}
NH_4^+	NH_3

$$HCl + H_2O \rightarrow H_3O^+ + Cl^- \qquad K_c = 5.55 \times 10^8$$

ACID   BASE   CONJUGATE ACID | CONJUGATE BASE

$$H_2O + NH_3 \rightleftharpoons NH_4^+ + OH^- \qquad K_c = 3.12 \times 10^{-8}$$

ACID

to be able to give protons, an acid must have a hydrogen atom which can be lost:
e.g. **HCl; H_2SO_4**

BASE

to be a base, a substance must have a lone pair which can become a sigma pair as it accepts a proton
e.g. **:NH_3**

STRONG AND WEAK ACIDS AND BASES

The equilibrium constants for the two reactions above are very different.

The first one, 5.5×10^8, tells us that the reaction lies well over to the right. That means that there are practically no hydrogen chloride molecules left because they have almost all become ions. This solution would conduct **strongly** and so the hydrochloric acid is described as a **strong acid.**

In the second reaction the constant is 3.12×10^{-8}. This tells us that about one in a thousand ammonia molecules have reacted with water. There are hardly any ions in this solution and so it conducts **weakly**. Ammonia is called a **weak base.**

So in water **strong acids** and **bases** exist mainly as **ions**, **weak acids** and **bases** exist mainly as **molecules.**

Common **strong acids** are:

HCl; H_2SO_4; HNO_3

Common **weak acids** are:

ethanoic, CH_3COOH; carbonic, H_2CO_3

Common **strong bases** are:

NaOH; KOH

Ammonia is a **weak base.**

Ampholytes

The water in the first reaction above is acting as a **base**. One of its lone pairs accepts a proton.
In the second reaction the water is being an **acid**, giving one of its protons to ammonia.

Substances like water, with hydrogens that can be lost and lone pairs that can be protonated, are called **ampholytes**. They can be either acids or bases depending on what they react with.

You can see some other common ampholytes in the list on the top right, because they appear in both the acid and base columns.

Even ammonia can be an ampholyte in the right conditions, having NH_4^+ as its conjugate acid and NH_2^- as its conjugate base.

Strength of acids

The strength of an acid (how well it protonates) is measured by the equilibrium constant for the reaction:

$$HA + H_2O \rightleftharpoons H_3O^+ + A^-$$

$K_c = \dfrac{[H_3O^+] \times [A^-]}{[HA] \times [H_2O]}$

$K_a = K_c \times 55.5 = \dfrac{[H_3O^+] \times [A^-]}{[HA]}$

$[H_2O]$ is nearly constant at 55.5 mol dm^{-3} in most bench solutions, so this expression can be simplified

K_a is called the **acidity** or **dissociation constant** for the acid.

If K_a is greater than 10^2 the acid is strong, while if it is less than 10^{-2} the acid is weak.

For an acid HA:

$$HA + H_2O \rightleftharpoons H_3O^+ + A^-$$

$K_a = \dfrac{[H_3O^+] [A^-]}{[HA]}$

In the same way, it is possible to write a K_b expression for a base and a K_w expression for water. Compare the three:

For the conjugate base: A$^-$:

$$A^- + H_2O \rightleftharpoons HA^+ + OH^-$$

$K_b = \dfrac{[HA][OH^-]}{[A^-]}$

For water:

$$H_2O + H_2O \rightleftharpoons H_3O^+ + OH^-$$

$K_w = [H_3O^+][OH^-]$

Now look what happens when we multiply the first two together and cancel out terms on the top and bottom lines:

$K_a \times K_b = \dfrac{[H_3O^+][\cancel{A^-}]}{[\cancel{HA}]} \times \dfrac{[\cancel{HA}][OH^-]}{[\cancel{A^-}]}$

$= [H_3O^+][OH^-] = K_w$

K_w is called the **dissociation constant** for water and has the value 1×10^{-14} mol^2 dm^{-6} at 25 °C.

WHAT CONTROLS THE STRENGTH OF AN ACID?

An energy cycle helps us focus on the two key ideas that help us answer this question:

1. How easy is it to break the bond joining the hydrogen to the rest of the acid particle? If the bond is weak, then the acid is likely to be strong.

2. How stable is the anion formed? If the anion is stable, the acid is likely to be strong.

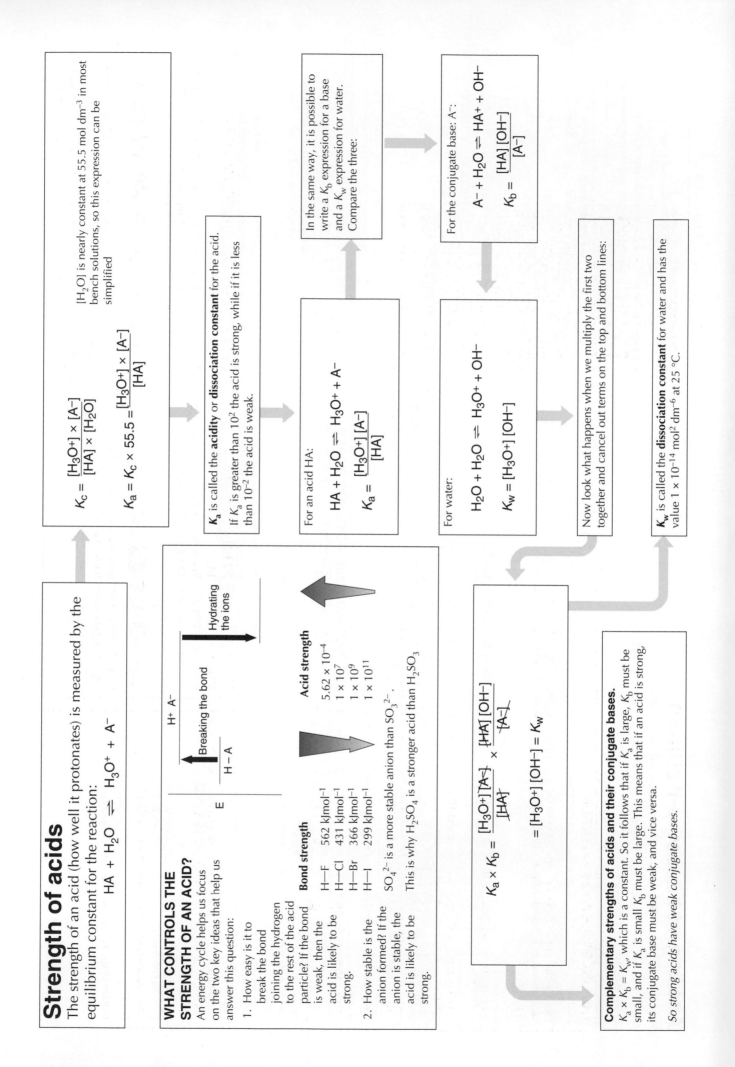

E

Breaking the bond

H — A

H$^+$ A$^-$

Hydrating the ions

Bond strength

H—F	562 kJmol^{-1}
H—Cl	431 kJmol^{-1}
H—Br	366 kJmol^{-1}
H—I	299 kJmol^{-1}

Acid strength

5.62×10^{-4}
1×10^7
1×10^9
1×10^{11}

SO_4^{2-} is a more stable anion than SO_3^{2-}.

This is why H_2SO_4 is a stronger acid than H_2SO_3

Complementary strengths of acids and their conjugate bases.

$K_a \times K_b = K_w$, which is a constant. So it follows that if K_a is large, K_b must be small, and if K_a is small K_b must be large. This means that if an acid is strong, its conjugate base must be weak, and vice versa.

So strong acids have weak conjugate bases.

The log scale and p-notation

Scientific calculator

Converting in and out of the p-notation is very easy with a calculator using the log and 10^x buttons; you do it like this:

THE LOG SCALE AND P-NOTATION

Numbers can be expressed on more than one scale.

On a normal number line, the gaps between the numbers are equal. For example, on this number line, the distance between 10 and 20 is the same as the distance between 20 and 30 and so on.

$$-50 \quad -40 \quad -30 \quad -20 \quad -10 \quad 0 \quad 10 \quad 20 \quad 30 \quad 40 \quad 50 \quad 60 \quad 70 \quad 80$$

On a **log scale** the distance between 1 and 10 and between 10 and 100 (in other words between powers of 10) is kept the same.

0.00001	0.0001	0.001	0.01	0.1	1	10	100	1000	10000	100000			
10^{-5}	10^{-4}	10^{-3}	10^{-2}	10^{-1}	10^0	10^1	10^2	10^3	10^4	10^5	10^6	10^7	10^8

The **p-notation** is a modified log scale used by chemists to express small values simply.

log scale													
10^{-5}	10^{-4}	10^{-3}	10^{-2}	10^{-1}	10^0	10^1	10^2	10^3	10^4	10^5	10^6	10^7	10^8
5	4	3	2	1	0	−1	−2	−3	−4	−5	−6	−7	−8

p − notation

pH

The **p-notation**, as applied to the hydrogen ion concentration, is formally defined as minus $\log_{10}$ of the hydrogen concentration, or mathematically as:

$$pH = -\log_{10}[H_3O^+] \text{ or } [H_3O^+] = 10^{-pH}$$

The notation is also very often applied to $[OH^-]$; K_a; K_b; and K_w, so

$$pOH = -\log_{10}[OH^-] \text{ and } pK_a = -\log_{10}K_a$$
and $pK_b = -\log_{10}K_b$, etc.

FROM STANDARD NUMERICAL FORM TO P-NOTATION:

e.g. if K_a for ethanoic acid is 1.8×10^{-5} mol dm^{-3}, pK_a for the acid can be found remembering that p$K_a = -\log_{10}K_a$ and pressing the following calculator keys:

[1] [.] [8] [EXP] [5] [+/−] [log] [+/−]

If you do this you will get the value 4.74, so pK_a for ethanoic acid is 4.74.

Similarly, if K_w for water is 1×10^{-14}, then pK_w is 14, and if $[H_3O^+] = 1 \times 10^{-7}$, then the pH of the solution will be 7.

FROM P-NOTATION INTO STANDARD NUMERICAL FORM:

e.g. if pH = 3.6 the hydrogen ion concentration can be found by using the relation $[H_3O^+] = 10^{-pH}$ and pressing the following keys:

[3] [.] [6] [+/−] [10^x]

If you do this you will get the value 2.51×10^{-4} and don't forget the units or you'll lose marks!

Calculations involving pH and pOH in strong acids and bases

EXAMPLES IN WHICH ONLY ONE PROTON IS TRANSFERRED PER MOLECULE

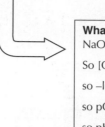

What is the pH of 2 mol dm⁻³ HCl?
HCl is a strong acid, so it exists as ions.

So $[H_3O^+] = [HCl] = 2$ mol dm⁻³,

so pH $= -\log_{10}[H_3O^+] = -\log_{10}2$,

so pH $= -0.3$.

What is the pH of 2 mol dm⁻³ NaOH?
NaOH is a strong base, so it exists as ions.

So $[OH^-] = [NaOH] = 2$ mol dm⁻³,

so $-\log_{10}[OH^-] = -\log_{10}2$,

so pOH $= -0.3$,

so pH $= 14 - -0.3 = 14.3$.

EXAMPLES IN WHICH TWO PROTONS ARE TRANSFERRED PER MOLECULE

What is the pH of 2 mol dm⁻³ H_2SO_4?
H_2SO_4 is a strong dibasic acid. It has two protons to give to the water.

So $[H_3O^+] = 2 \times [H_2SO_4] = 2 \times 2$ mol dm⁻³,

so pH $= -\log_{10} 4 = -0.6$.

What is the pH of 0.25 mol dm⁻³ $Ca(OH)_2$?
$Ca(OH)_2$ is a strong di-acidic base. It can receive two protons from the water.

So $[OH^-] = 2 \times [Ca(OH)_2] = 2 \times 0.25$ mol dm⁻³,

so pOH $= -\log_{10} 0.5 = 0.3$,

so pH $= 14 - 0.3 = 13.7$.

EXAMPLES IN WHICH THE CONCENTRATIONS ARE GIVEN IN STANDARD FORM

What is the pH of 1×10^{-3} mol dm⁻³ HNO_3?
HNO_3 is another strong acid.

So $[H_3O^+] = [HNO_3] = 1 \times 10^{-3}$ mol dm⁻³,

so pH $= -\log_{10}(1 \times 10^{-3}) = 3$.

What is the pH of 1×10^{-3} mol dm⁻³ KOH?
KOH is another strong base.

So $[OH^-] = [KOH] = 1 \times 10^{-3}$ mol dm⁻³,

so pOH $= -\log_{10}(1 \times 10^{-3}) = 3$,

so pH $= 14 - 3 = 11$.

AN EXAMPLE OF A VERY DILUTE SOLUTION WHERE THE HYDROXONIUM IONS, H_3O^+, FROM WATER CANNOT BE IGNORED

What is the pH of 1×10^{-8} mol dm⁻³ HNO_3?
Following the pattern of the example above you might think that the answer would be 8. But if you think about it, this cannot be possible: a pH of 8 is alkaline, and however dilute this acid is it will not give a pH greater than 7.

The reason for this wrong answer is that in the calculations above we have ignored the H_3O^+ from water itself because it has no measurable effect in a strong acid. Here, however, the acid is so dilute that the H_3O^+ from water becomes significant.

So here there are 1×10^{-7} mol dm⁻³ H_3O^+ from water and 1×10^{-8} mol dm⁻³ H_3O^+ from the acid. The equilibrium position in the water will change slightly, but, approximating, we can say that this is the same as 1.1×10^{-7} mol dm⁻³ so
pH $= -\log_{10}(1.1 \times 10^{-7}) = 6.96$.

MIXTURES OF ACIDS AND BASES

What is the pH of a mixture of 50 cm³ of 0.1 mol dm⁻³ HCl and 30 cm³ of 0.1 mol dm⁻³ NaOH?
Both the acid and the base are of the same concentration, so the 30 cm³ of base will have neutralized 30 cm³ of the acid.

This leaves 20 cm³ of acid unreacted, but this acid is now in a volume of 80 cm³, so the remaining acid will be diluted 4 times giving a concentration of 0.025 mol dm⁻³.

As this is a strong acid, the $[H_3O^+] = 0.025$ mol dm⁻³ so the pH $= -\log_{10} 0.025 = 1.60$

Weak acid calculations

For any weak acid we can write the general equilibrium statement

$$HA \rightleftharpoons H^+ + A^-$$

So if the acid is weak, the concentration of HA after splitting up is nearly the same as it was before. Weak acids exist mainly as molecules. For example only about 1 in a thousand ethanoic acid molecules react with water, so if the original [HA] was, say, 0.100 mol dm^{-3}, then after reacting with water [HA] would be 0.099 mol dm^{-3}.

If we ignore the H$^+$ contribution from the water, [H$^+$] must equal [A$^-$] because every time an acid molecule splits up it gives one A$^-$ for every one H$^+$.

Applying the equilibrium law, we get:

$$K_a = \frac{[H^+][A^-]}{[HA]}$$

which is the same as:

dissociation constant of the acid $\longrightarrow$ $K_a = \dfrac{[H^+]^2}{[HA]}$ because as we said above [H$^+$] = [A$^-$]

related to the pH

$\approx$ original concentration of the acid

In all the weak acid calculations you will be given values of two of K_a, pH (hence [H$^+$]), or [HA] and asked to calculate the third. You do this using the equation above.

Here are some examples:

The K_a for ethanoic acid is 1.8 × 10^{-5} mol dm^{-3}. Find the pH of a 0.1 mol dm^{-3} solution.

So we know K_a and [HA]. If we can find [H$^+$], then we can convert to the p-notation easily.

Rearranging the equation above, and substituting in, we get:

$$
\begin{aligned}
[H^+]^2 &= K_a \times [HA] \\
&= 1.8 \times 10^{-5} \times 0.1 \\
&= 1.8 \times 10^{-6}
\end{aligned}
$$

$$
\begin{aligned}
[H^+] &= \sqrt{1.8 \times 10^{-6}} \\
&= 1.34 \times 10^{-3}
\end{aligned}
$$

pH = 2.87

The pH of a 0.01 mol dm^{-3} solution is 4.1. Find the K_a of the acid.

Here we do not have to rearrange the equation because it is K_a that we want.

$$
\begin{aligned}
K_a &= \frac{[H^+]^2}{[HA]} \\
&= \frac{10^{-4.1} \times 10^{-4.1}}{0.01} \\
&= 6.3 \times 10^{-7} \text{ mol dm}^{-3}
\end{aligned}
$$

An acid whose K_a is 4.3 × 10^{-6} has a pH of 4.7. Find the concentration of the acid

In this example we must again rearrange the equation so that we get:

$$
\begin{aligned}
[HA] &= \frac{[H^+]^2}{K_a} \\
&= \frac{10^{-4.7} \times 10^{-4.7}}{4.3 \times 10^{-6}} \\
&= 9.25 \times 10^{-5} \text{mol dm}^{-3}
\end{aligned}
$$

Buffer solutions
resist changes in pH on addition of acid or alkali

Buffer solutions tend to resist changes in pH when small amounts of acid or base are added and their pH is not affected by dilution.

Acidic buffers are made of a solution of a weak acid and its conjugate base, e.g. a solution of ethanoic acid and sodium ethanoate.

Basic buffers are made of a solution of a weak base and its conjugate acid, e.g. a solution of ammonia and ammonium chloride.

A weak acid like this exists mainly as molecules. These molecules can provide more hydroxonium ions if the existing ones are removed by adding a base.

This salt contains ethanoate ions, the conjugate base of ethanoic acid. If acid is added they can react with it.

If the two are added together a buffer solution is made.

Ethanoic acid

+

Sodium ethanoate

=

100 ml

Buffer solution

BUFFER CALCULATIONS
For any weak acid:

$$HA + H_2O \rightleftharpoons H_3O^+ + A^-$$

applying the equilibrium law:

$$K_a = [H_3O^+] \times \frac{[A^-]}{[HA]}$$

or rearranging the equation:

$$\frac{K_a}{[H_3O^+]} = \frac{[A^-]}{[HA]}$$

The ratio of concentration of conjugate base to acid controls the ratio of K_a to $[H_3O^+]$. Dilution does not affect the value of this ratio — both concentrations are changed equally — so the pH of a buffer is not affected by dilution.

Example 1
What is the pH of a buffer made by adding 0.2 mol of sodium ethanoate to 500 cm³ of 0.1 mol dm⁻³ ethanoic acid, given that K_a for the acid is 1.8×10^{-5}?

$$\frac{K_a}{[H_3O^+]} = \frac{0.4}{0.1}$$

so $[H_3O^+] = K_a \div 4$
$= 4.5 \times 10^{-6}$

and pH = 5.35

Example 2
How can you make a buffer of pH 4.5 from propanoic acid, pK_a = 4.87?

$$\frac{K_a}{[H_3O^+]} = \frac{[A^-]}{[HA]}$$

$$\frac{10^{-4.87}}{10^{-4.5}} = \frac{1.35 \times 10^{-5}}{3.16 \times 10^{-5}}$$

$$\frac{0.74}{1.0} = \frac{[A^-]}{[HA]}$$

so take 1 dm⁻³ of 1 mol dm⁻³ propanoic acid and add 0.43 mol (0.43 × 96 g = 42.3 g) of sodium propanoate.

USES OF BUFFERS
Buffers are vital in almost all biological systems where a change in pH can have a great effect on the functioning of a cell. To prevent this, all injections and eye drops, for example, are buffered.

Buffers are also important in industry. Both in the dyeing and electroplating industries, maintaining the pH of the process is essential to its success.

HOW BUFFERS WORK
We have seen that a buffer contains particles which can react with either any acid or any base which is added.

$$CH_3COOH + H_2O \rightleftharpoons CH_3COO^- + H_3O^+$$

lots of spare acid molecules because ethanoic acid is a weak acid

lots of spare conjugate base because sodium ethanoate has been added to the acid solution

Addition of a base
Any base that is added reacts with the hydroxonium ions, H_3O^+, and uses them up. This means that there is no back reaction in the equilibrium, but the forward reaction goes on with more ethanoic acid molecules protonating water and replacing most of the hydroxonium ions that were removed by the base. So the pH hardly changes at all.

Addition of acid
Adding acid, that is more hydroxonium ions, H_3O^+, increases the rate of the back reaction as ethanoate ions collide more often with acid ions until most of the extra ones are removed. Once again the pH hardly changes.

Hydration and hydrolysis: reactions with water

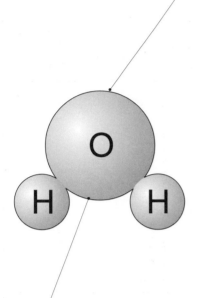

HYDRATION

When any substance is dissolved in water, the particles in it become surrounded by water molecules. This process of surrounding the particles with water is called hydration and the ions formed are called aquo-ions. The equations below represent these changes for various substances.

e.g.

sodium chloride	$NaCl(s) + H_2O(l) \rightarrow Na^+(aq) + Cl^-(aq)$
carbon dioxide	$CO_2(g) + H_2O(l) \rightarrow CO_2(aq)$
ammonia	$NH_3(g) + H_2O(l) \rightarrow NH_3(aq)$
aluminium chloride	$AlCl_3(s) + H_2O(l) \rightarrow Al^{3+}(aq) + 3Cl^-(aq)$
sodium carbonate	$Na_2CO_3(s) + H_2O(l) \rightarrow 2Na^+(aq) + CO_3^{2-}(aq)$

Complex ions

The water molecules around ions, for example a sodium or chloride ion, are fairly weakly attracted and so are constantly changing as other water molecules replace those nearest the ion. However, some ions are sufficiently small and highly charged to attract the water molecules (or other particles in solution) so strongly that they become firmly joined to the ion, at least for a time. The particle formed is called a complex ion and the water or other surrounding particles are called **ligands**.

A modified equilibrium law expression can be written for this process and the equilibrium constant is called the stability constant, K_{stab}.

HYDROLYSIS

Once the particles are surrounded by water, they may react with it. The reaction of particles with the water is called hydrolysis. Hydrolysis reactions always lead to a change in pH of the solution because the relative number of hydroxonium or hydroxide ions is changed.

Covalent substances

(i) **with a positive reactive site** react with water making it acidic

positive reactive site $\delta+$ $\delta-$ $\rightarrow$ H_2CO_3 carbonic acid then $H_2CO_3(aq) + H_2O(l) \rightleftharpoons H_3O^+(aq) + HCO_3^-(aq)$

(ii) **with a prominent lone pair** react with water making it alkaline

$$\rightleftharpoons \quad H\cdots N - H^{\oplus} \quad + \quad :OH^-(aq)$$

Ionic substances

(i) **with small, highly charged cations** react making the water acidic

$$[Al(H_2O)_6]^{3+}(aq) + H_2O(l) \rightleftharpoons [Al(H_2O)_5OH]^{2+}(aq) + H_3O^+(aq)$$

Nearly all transition metal cations are hydrolysed by water making an acidic solution. The reason for this is that the d electrons do not shield the nucleus well so producing an ion with high polarising power.

(ii) **with anions which are the conjugate bases of weak acids** react making the water alkaline

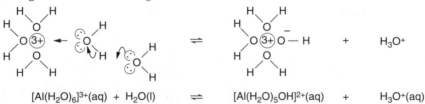

$$CO_3^{2-}(aq) + H_2O(l) \rightleftharpoons HCO_3^-(aq) + OH^-(aq)$$

Indicators are substances which change colour as the pH changes

An **indicator** is a weak acid whose conjugate base is a different colour.

$$HInd \rightleftharpoons H_3O^+ + Ind^-$$

colour A   colour B

In *acid* there will be a lot of H_3O^+ and so the equilibrium will lie to the left. The colour seen will be colour A.

In *alkaline* conditions, all the H_3O^+ ions will be used up and the equilibrium will move to the right. The colour seen will be colour B.

When there are equal amounts of the two colours, [HInd] will equal [Ind−] and the colour will be a mixture of A and B. This happens when pH = pK_{Ind} as shown below.

$$K_{Ind} = [H_3O^+] \times \frac{[Ind^-]}{[HInd]} \quad \text{when these two are equal} \quad \frac{[Ind^-]}{[HInd]} = 1$$

$$\text{so } K_{Ind} = [H_3O^+] \text{ and } pK_{Ind} = pH$$

The sudden change in colour seen in a titration is called the **end-point**. Some people are colour blind, but most people can detect a change in colour when the ratio of A to B is 1 : 10 or 10 : 1. This means that most indicators have a useful pH range of about 2 pH units, about one either side of the pK_{Ind} value as the table shows.

The **equivalence point** of the titration, when amounts of acid and base exactly balance, and the end point will be very close to each other if the correct indicator has been chosen.

Indicator	pK_{Ind}	Useful pH range	Colour change
Methyl orange	3.7	3.1 – 4.4	red to yellow
Bromophenol blue	4.0	3.0 – 4.6	yellow to blue
Methyl red	5.1	4.2 – 6.3	red to yellow
Bromothymol blue	7.0	6.0 – 7.6	yellow to blue
Phenolphthalein	9.3	8.3 – 10.0	colourless to red

INDICATOR STRUCTURE

Indicators are quite complicated molecules whose electronic structure changes as they are protonated or deprotonated. This change in electronic structure results in a change in electronic energy levels and so a change in colour when light falls on to the molecules and moves electrons between the energy levels.

Red methyl orange: protonated on this nitrogen: seen in acid solutions

Yellow methyl orange: deprotonated here: seen in alkaline conditions

TITRATIONS WITHOUT INDICATORS

If one of the reactants is coloured, an indicator is not necessary. In particular, redox titrations such as those involving the reduction of manganate (VII) ions or iodine can be done without an indicator.

For manganate (VII) titrations the colour change is between purple and colourless. Dilute sulphuric acid is always added because MnO_4^- is a better oxidising agent in acid conditions and the end point occurs when the colour changes from purple to colourless or from colourless to a faint pink.

Iodine which is brown in solution is titrated against thiosulphate solution. Here the end point occurs when the solution goes from pale yellow to colourless. This change can be made more obvious by adding some starch solution once the iodine has reached a straw colour. The starch forms a blue compound with the iodine which disappears as the iodine is used up.

thiosulphate

Add starch

brown iodine

iodine pale straw colour

iodine now dark blue

iodine used up solution colourless

Titration curves I

Titration curves are graphs showing how the pH of an acid or base changes as it is neutralised. The curve is usually obtained by placing a glass electrode into the titration flask and recording the pH as the alkali is run in.

STRONG ACID NEUTRALISED BY STRONG BASE

This curve shows what happens when 1 mol dm^{-3} sodium hydroxide is added to 50 cm^3 of 1 mol dm^{-3} hydrochloric acid.

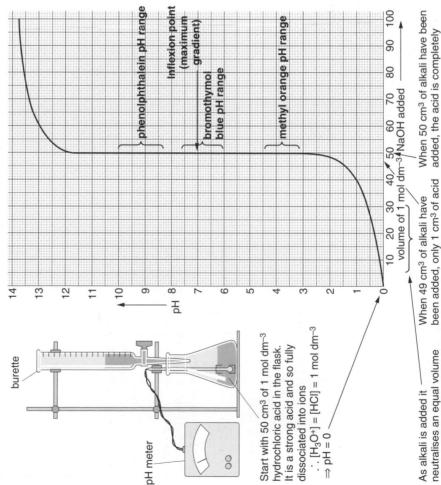

phenolphthalein pH range

Inflexion point (maximum gradient)

bromothymol blue pH range

methyl orange pH range

burette

pH meter

Start with 50 cm^3 of 1 mol dm^{-3} hydrochloric acid in the flask. It is a strong acid and so fully dissociated into ions
$$\therefore [H_3O^+] = [HCl] = 1 \text{ mol dm}^{-3}$$
$$\Rightarrow pH = 0$$

As alkali is added it neutralises an equal volume of acid making salt and water. The remaining acid is diluted by the increased volume of water so the concentration of hydrogen ions decreases and the pH increases

When 49 cm^3 of alkali have been added, only 1 cm^3 of acid remains un-neutralised. This 1 cm^3 of acid is in 99 cm^3 of solution so it is diluted about 100 times.
$$[H_3O^+] = \frac{1}{100} = 10^{-2}$$
$$\therefore pH = 2$$

When 50 cm^3 of alkali have been added, the acid is completely reacted leaving salt and water. The pH is 7. This is indicated by the inflexion point. From now on, alkali added has nothing to react with so the pH rises towards a final value of 14

WEAK ACID NEUTRALISED BY STRONG BASE

This curve shows what happens when 1 mol dm^{-3} sodium hydroxide is added to 50 cm^3 of 1 mol dm^{-3} ethanoic acid.

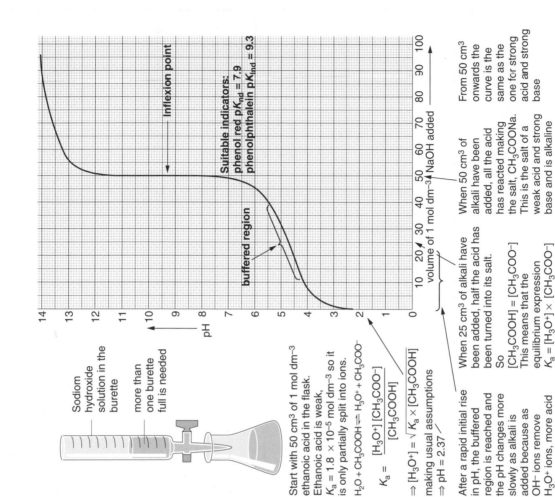

Inflexion point

Suitable indicators:
phenol red $pK_{ind} = 7.9$
phenolphthalein $pK_{ind} = 9.3$

buffered region

Sodium hydroxide solution in the burette

more than one burette full is needed

Start with 50 cm^3 of 1 mol dm^{-3} ethanoic acid in the flask. Ethanoic acid is weak, $K_a = 1.8 \times 10^{-5}$ mol dm^{-3} so it is only partially split into ions.
$$H_2O + CH_3COOH \rightleftharpoons H_3O^+ + CH_3COO^-$$
$$K_a = \frac{[H_3O^+][CH_3COO^-]}{[CH_3COOH]}$$
$$\Rightarrow [H_3O^+] = \sqrt{K_a \times [CH_3COOH]}$$
making usual assumptions
$$\Rightarrow pH = 2.37$$

After a rapid initial rise in pH, the buffered region is reached and the pH changes more slowly as alkali is added because as OH$^-$ ions remove H$_3$O$^+$ ions, more acid splits up producing new H$_3$O$^+$ ions

When 25 cm^3 of alkali have been added, half the acid has been turned into its salt.
So
$$[CH_3COOH] = [CH_3COO^-]$$
This means that the equilibrium expression
$$K_a = [H_3O^+] \times \frac{[CH_3COO^-]}{[CH_3COOH]}$$
becomes $K_a = [H_3O^+]$
or $pK_a = pH$

When 50 cm^3 of alkali have been added, all the acid has reacted making the salt, CH$_3$COONa. This is the salt of a weak acid and strong base and is alkaline as a result of hydrolysis. The inflexion point is at 9.24 not 7

From 50 cm^3 onwards the curve is the same as the one for strong acid and strong base

Titration curves II

STRONG ACID NEUTRALISED BY WEAK BASE

This curve shows what happens when 1 mol dm⁻³ ammonia is added to 50 cm³ of 1 mol dm⁻³ hydrochloric acid.

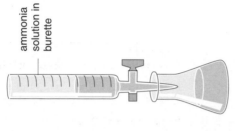

ammonia solution in burette

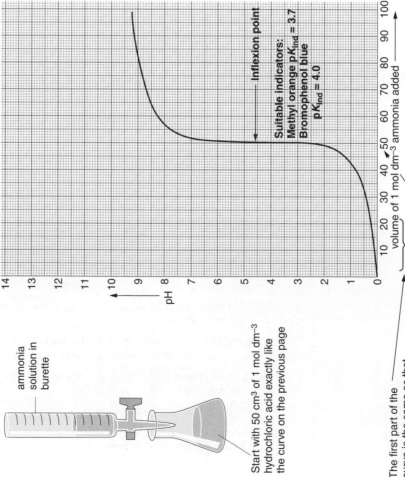

pH

Inflexion point

Suitable indicators:
Methyl orange pK_{ind} = 3.7
Bromophenol blue
pK_{ind} = 4.0

volume of 1 mol dm⁻³ ammonia added

Start with 50 cm³ of 1 mol dm⁻³ hydrochloric acid exactly like the curve on the previous page

The first part of the curve is the same as that for strong acid against strong base

When 50 cm³ of alkali have been added all the acid has reacted making the salt NH₄Cl. This is the salt of a strong acid and a weak base and is acidic due to the hydrolysis of the ammonium ion so the pH of the inflexion point is 4.6

$$NH_4^+(aq) + H_2O(l) \rightleftharpoons H_3O^+(aq) + NH_3(aq)$$

From 50 cm³ onwards there is no acid for the ammonia to react with, but because ammonia is a weak base and exists mainly as molecules the pH rises only towards 10 not 14 as with strong base.

$$NH_3(aq) + H_2O(l) \rightleftharpoons NH_4^+(aq) + OH^-(aq)$$
99.9%   0.1%

WEAK ACID NEUTRALISED BY WEAK BASE

This curve shows what happens when 1 mol dm⁻³ ammonia is added to 50 cm³ of 1 mol dm⁻³ ethanoic acid.

ammonia solution in burette

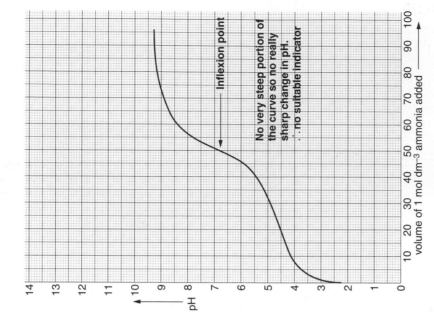

pH

Inflexion point

No very steep portion of the curve so no really sharp change in pH.
∴ no suitable indicator

volume of 1 mol dm⁻³ ammonia added

Start with 50 cm³ of 1 mol dm⁻³ ethanoic acid in the flask like the previous weak acid curve. Find the inflexion point by examining the pH curve. No suitable indicator

The first part of the curve is like the first part of the weak acid/strong base curve. The last part of the curve is like the last part of the strong acid/weak base curve. The central part with the inflexion point is unlike all the other curves in that it does not have an almost vertical section. This means the pH never changes rapidly enough to make any indicator give a sharp colour change.

Redox reactions

Reduction and oxidation reactions involve
the transfer or loss and gain of electrons.

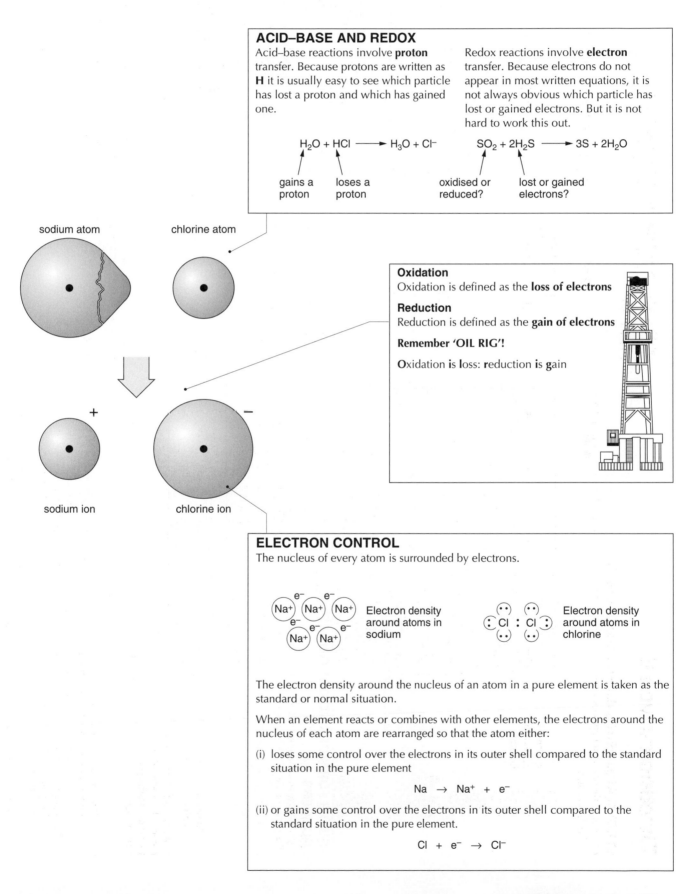

ACID–BASE AND REDOX

Acid–base reactions involve **proton** transfer. Because protons are written as **H** it is usually easy to see which particle has lost a proton and which has gained one.

Redox reactions involve **electron** transfer. Because electrons do not appear in most written equations, it is not always obvious which particle has lost or gained electrons. But it is not hard to work this out.

$$H_2O + HCl \longrightarrow H_3O + Cl^-$$

gains a proton   loses a proton

$$SO_2 + 2H_2S \longrightarrow 3S + 2H_2O$$

oxidised or reduced?   lost or gained electrons?

sodium atom   chlorine atom

sodium ion   chlorine ion

Oxidation

Oxidation is defined as the **loss of electrons**

Reduction

Reduction is defined as the **gain of electrons**

Remember 'OIL RIG'!

Oxidation **i**s **l**oss: **r**eduction **i**s **g**ain

ELECTRON CONTROL

The nucleus of every atom is surrounded by electrons.

Electron density around atoms in sodium

Electron density around atoms in chlorine

The electron density around the nucleus of an atom in a pure element is taken as the standard or normal situation.

When an element reacts or combines with other elements, the electrons around the nucleus of each atom are rearranged so that the atom either:

(i) loses some control over the electrons in its outer shell compared to the standard situation in the pure element

$$Na \rightarrow Na^+ + e^-$$

(ii) or gains some control over the electrons in its outer shell compared to the standard situation in the pure element.

$$Cl + e^- \rightarrow Cl^-$$

Oxidation numbers

The electron density around an atom in a substance is shown by **oxidation numbers**. These numbers indicate the electron density around an atom *compared to its situation in the pure element.*

carbon has lost some control over 4 electrons so oxidation number is + IV

each oxygen has gained some control over 2 electrons so oxidation number is – II

AN OXIDATION NUMBER IS MADE OF TWO PARTS:

(i) the **sign**

- if the sign is *positive*, the atom has *lost* control of electrons
- if the sign is *negative*, the atom has *gained* control of electrons

(ii) the **number** is always written as a Roman number

- this gives the *number* of electrons over which electron control has changed compared to the situation in the pure element

RULES FOR WORKING OUT OXIDATION NUMBERS

1. The oxidation number of atoms in a pure element is zero.

0	0	0	0
Mg	Fe	Cl_2	H_2	O_2

2. In hydrogen compounds, the oxidation number of hydrogen is always +I, except in metal hydrides where it is –I.

+I	+I	+I	+I
HCl	CH_4	H_2O	H_2SO_4

3. In oxygen compounds the oxidation number of oxygen is always –II, except in peroxides where it is –I and in fluorine compounds where it is +II.

–II	–II	–II	–I	+II
CO_2	H_2O	$KMnO_4$	BUT	H_2O_2 and F_2O

4. The oxidation number of an ion made from a single atom equals the value of the charge on the ion.

+I	+II	+III	–II	–I
Na^+	Mg^{2+}	Fe^{3+}	O^{2-}	Br^-

5. The sum of all the oxidation numbers of the atoms in the formula of a compound is zero.

CH_4 –IV + 4(I) = 0 : H_2SO_4 $2 \times (I) + VI + 4(II) = 0$

6. The sum of all the oxidation numbers of the atoms in the formula of a molecular ion equals the charge on the ion.

SO_4^{2-} +VI + 4(–II) = –2 : NO_3^- +V + 3(–II) = –1

RECOGNISING REDOX EQUATIONS

Redox equations are recognised by:

(i) working out all the oxidation numbers of the atoms in the equation;

(ii) seeing if the oxidation number of any atom has changed. If it has the reaction is a redox one.

+II –II	+I –I	+II –I	+I –II
PbO	+ 2HCl	→	$PbCl_2$ + H_2O

No change → not redox

+IV –II	+I –I	+II –I	+I –II	0
PbO_2	+ 4HCl	→	$PbCl_2$ +	$2H_2O$ + Cl_2

Oxidation numbers → this is redox
do change

Recognising reduction and oxidation

A change in the oxidation number of an atom during a reaction means that the atom has either lost or gained electron control.

REDUCTION
If the oxidation number of an atom has become more **negative**, control over electrons has been gained. The **substance** containing that atom has been **reduced**.

OXIDATION
If the oxidation number of an atom has become more **positive**, control over electrons has been lost. The **substance** containing that atom has been **oxidised**.

REDOX IS A TWO-WAY PROCESS
Electron control lost by one atom is gained by another. For this reason, reduction and oxidation always happen at the same time. This is why the term **redox** is usually used. If the words **oxidation** or **reduction** are used it is because we are only focusing on one reactant.

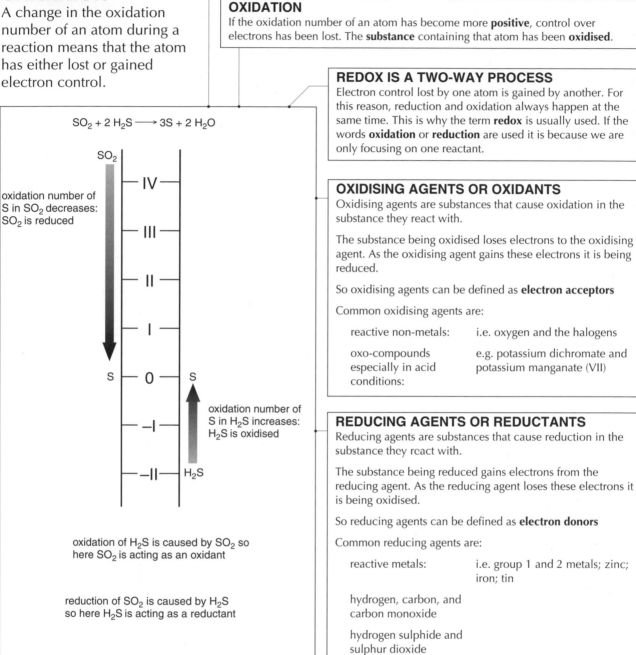

$$SO_2 + 2 H_2S \longrightarrow 3S + 2 H_2O$$

oxidation number of S in SO_2 decreases: SO_2 is reduced

oxidation number of S in H_2S increases: H_2S is oxidised

oxidation of H_2S is caused by SO_2 so here SO_2 is acting as an oxidant

reduction of SO_2 is caused by H_2S so here H_2S is acting as a reductant

OXIDISING AGENTS OR OXIDANTS
Oxidising agents are substances that cause oxidation in the substance they react with.

The substance being oxidised loses electrons to the oxidising agent. As the oxidising agent gains these electrons it is being reduced.

So oxidising agents can be defined as **electron acceptors**

Common oxidising agents are:

reactive non-metals:	i.e. oxygen and the halogens
oxo-compounds especially in acid conditions:	e.g. potassium dichromate and potassium manganate (VII)

REDUCING AGENTS OR REDUCTANTS
Reducing agents are substances that cause reduction in the substance they react with.

The substance being reduced gains electrons from the reducing agent. As the reducing agent loses these electrons it is being oxidised.

So reducing agents can be defined as **electron donors**

Common reducing agents are:

reactive metals:	i.e. group 1 and 2 metals; zinc; iron; tin
hydrogen, carbon, and carbon monoxide	
hydrogen sulphide and sulphur dioxide	

TESTING FOR OXIDISING AND REDUCING AGENTS
Acids and bases are tested for using indicators. Indicators are weak acids whose conjugate base is a different colour.

Oxidising and reducing agents are tested for in a similar way by using a reducing or oxidising agent whose conjugate partner is a different colour.

Tests for reducing agents

Add the substance to	Look for the colour change		Oxidising agent	Conjugate reducing agent
acidified potassium manganate (VII) solution	purple	→ colourless	$MnO_4^-(aq)$	$Mn^{2+}(aq)$
acidified potassium dichromate solution	orange	→ green/blue	$Cr_2O_7^{2-}(aq)$	$Cr^{3+}(aq)$

Tests for oxidising agents

Add the substance to	Look for the colour change		Reducing agent	Conjugate oxidising agent
iron (II) sulphate solution	green	→ yellow	$Fe^{2+}(aq)$	$Fe^{3+}(aq)$
potassium iodide solution	colourless	→ yellow/brown	$I^-(aq)$	$I_2(aq)$

Conjugate redox pairs

	OXIDANT	REDUCTANT
Each oxidising agent has a reduced form called its conjugate reducing agent.	Cl_2	Cl^-
Each reducing agent has an oxidised form called its conjugate oxidising agent.	H_2O	H_2
Strong oxidising agents have weak conjugate reducing agents,	MnO_4^-	Mn^{2+}
	$Cr_2O_7^{2-}$	Cr^{3+}
while strong reducing agents have weak conjugate oxidising agents.	Mg^{2+}	Mg
	CO_2	CO

RELATIONSHIP BETWEEN CONJUGATE REDOX PAIRS AND A COMPARISON WITH ACID–BASE PAIRS

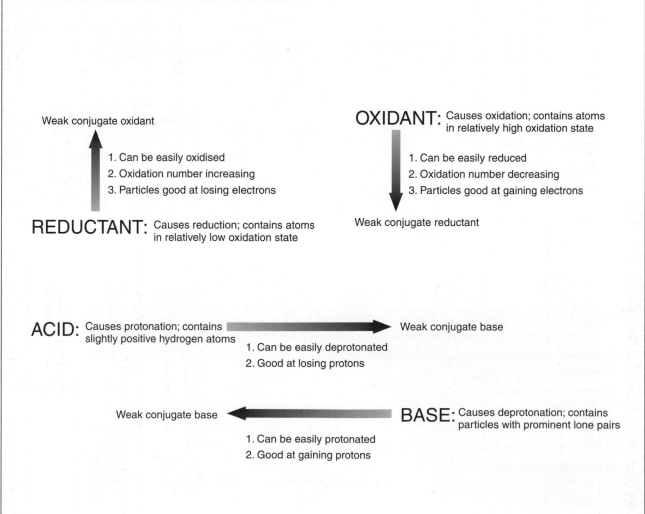

Weak conjugate oxidant

1. Can be easily oxidised
2. Oxidation number increasing
3. Particles good at losing electrons

REDUCTANT: Causes reduction; contains atoms in relatively low oxidation state

OXIDANT: Causes oxidation; contains atoms in relatively high oxidation state

1. Can be easily reduced
2. Oxidation number decreasing
3. Particles good at gaining electrons

Weak conjugate reductant

ACID: Causes protonation; contains slightly positive hydrogen atoms → Weak conjugate base

1. Can be easily deprotonated
2. Good at losing protons

Weak conjugate base ← BASE: Causes deprotonation; contains particles with prominent lone pairs

1. Can be easily protonated
2. Good at gaining protons

Standard electrode potentials

STANDARD ELECTRODE POTENTIALS

The change in electron density around atoms during oxidation and reduction means that there is a difference in electrical potential between the oxidised and reduced forms of the element.

in calcium metal the ions are surrounded by delocalised electrons

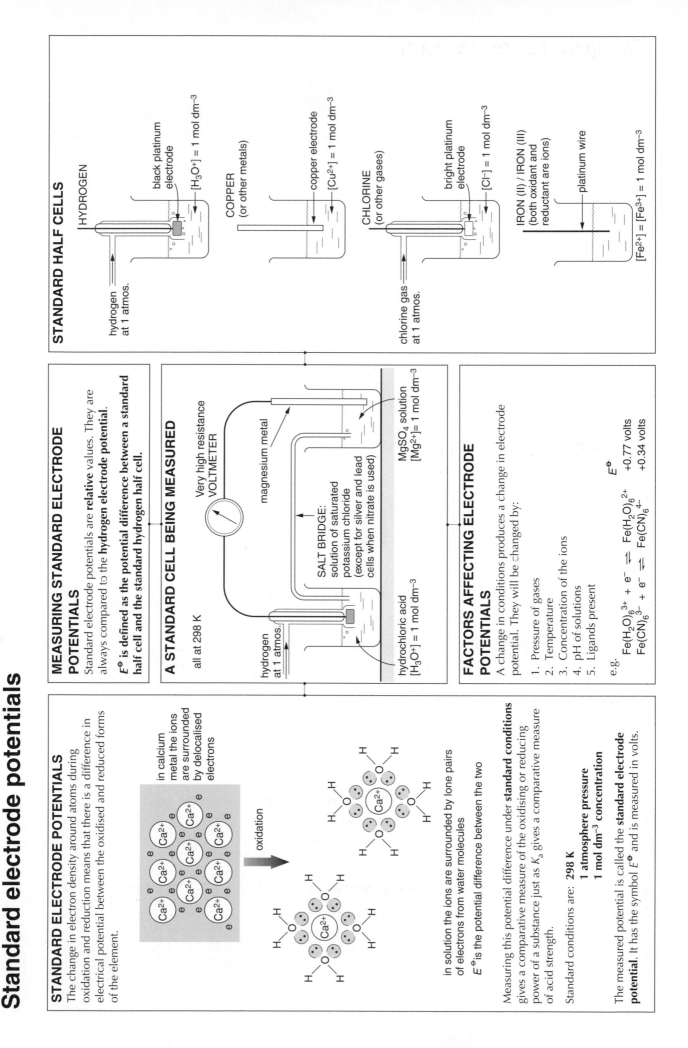

oxidation

in solution the ions are surrounded by lone pairs of electrons from water molecules

$E^{\ominus}$ is the potential difference between the two

Measuring this potential difference under **standard conditions** gives a comparative measure of the oxidising or reducing power of a substance just as K_a gives a comparative measure of acid strength.

Standard conditions are: **298 K**
1 atmosphere pressure
1 mol dm⁻³ concentration

The measured potential is called the **standard electrode potential**. It has the symbol $E^{\ominus}$ and is measured in volts.

MEASURING STANDARD ELECTRODE POTENTIALS

Standard electrode potentials are **relative** values. They are always compared to the **hydrogen electrode potential**.

$E^{\ominus}$ **is defined as the potential difference between a standard half cell and the standard hydrogen half cell.**

A STANDARD CELL BEING MEASURED

all at 298 K

Very high resistance VOLTMETER

magnesium metal

SALT BRIDGE: solution of saturated potassium chloride (except for silver and lead cells when nitrate is used)

$MgSO_4$ solution $[Mg^{2+}] = 1$ mol dm⁻³

hydrogen at 1 atmos.

hydrochloric acid $[H_3O^+] = 1$ mol dm⁻³

FACTORS AFFECTING ELECTRODE POTENTIALS

A change in conditions produces a change in electrode potential. They will be changed by:

1. Pressure of gases
2. Temperature
3. Concentration of the ions
4. pH of solutions
5. Ligands present

e.g.

$$Fe(H_2O)_6^{3+} + e^- \rightleftharpoons Fe(H_2O)_6^{2+} \qquad +0.77 \text{ volts}$$
$$Fe(CN)_6^{3-} + e^- \rightleftharpoons Fe(CN)_6^{4-} \qquad +0.34 \text{ volts}$$

$E^{\ominus}$

STANDARD HALF CELLS

HYDROGEN

black platinum electrode

$[H_3O^+] = 1$ mol dm⁻³

hydrogen at 1 atmos.

COPPER (or other metals)

copper electrode

$[Cu^{2+}] = 1$ mol dm⁻³

CHLORINE (or other gases)

bright platinum electrode

$[Cl^-] = 1$ mol dm⁻³

chlorine gas at 1 atmos.

IRON (II) / IRON (III) (both oxidant and reductant are ions)

platinum wire

$[Fe^{2+}] = [Fe^{3+}] = 1$ mol dm⁻³

Predicting redox reactions

The value of the standard electrode potential, $E^{\ominus}$, for a half cell gives a measure of the relative stability of the conjugate oxidising and reducing agents.

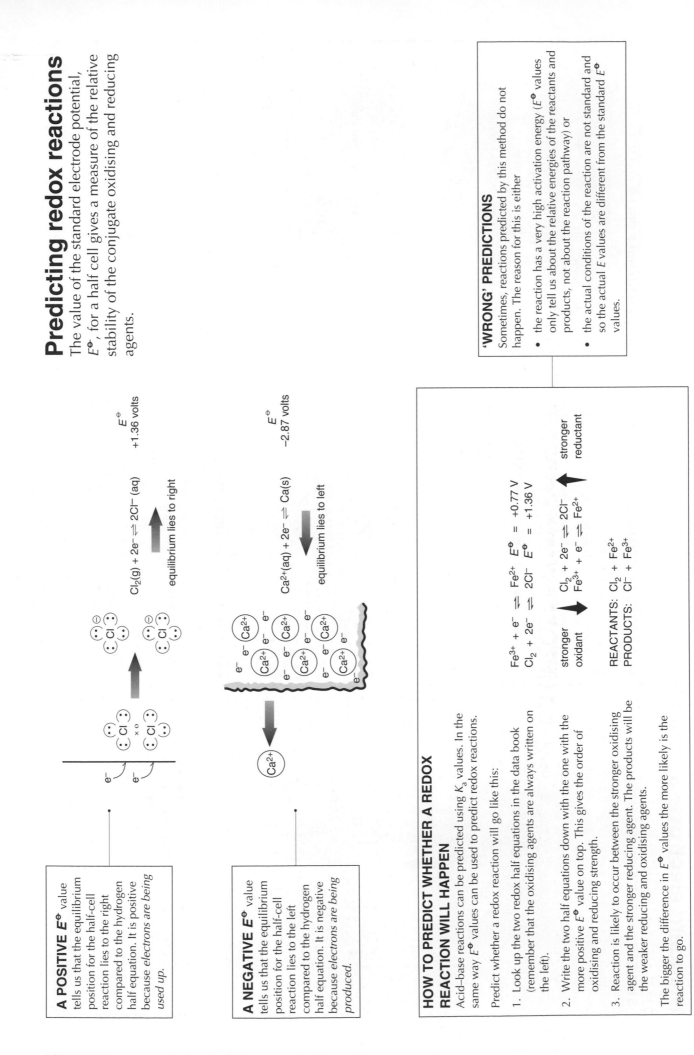

A POSITIVE $E^{\ominus}$ value
tells us that the equilibrium position for the half-cell reaction lies to the right compared to the hydrogen half equation. It is positive because *electrons are being used up.*

$$Cl_2(g) + 2e^- \rightleftharpoons 2Cl^-(aq) \qquad E^{\ominus} \quad +1.36 \text{ volts}$$

equilibrium lies to right

A NEGATIVE $E^{\ominus}$ value
tells us that the equilibrium position for the half-cell reaction lies to the left compared to the hydrogen half equation. It is negative because *electrons are being produced.*

$$Ca^{2+}(aq) + 2e^- \rightleftharpoons Ca(s) \qquad E^{\ominus} \quad -2.87 \text{ volts}$$

equilibrium lies to left

HOW TO PREDICT WHETHER A REDOX REACTION WILL HAPPEN

Acid–base reactions can be predicted using K_a values. In the same way $E^{\ominus}$ values can be used to predict redox reactions.

Predict whether a redox reaction will go like this:

1. Look up the two redox half equations in the data book (remember that the oxidising agents are always written on the left).

2. Write the two half equations down with the one with the more positive $E^{\ominus}$ value on top. This gives the order of oxidising and reducing strength.

3. Reaction is likely to occur between the stronger oxidising agent and the stronger reducing agent. The products will be the weaker reducing and oxidising agents.

The bigger the difference in $E^{\ominus}$ values the more likely is the reaction to go.

$$Fe^{3+} + e^- \rightleftharpoons Fe^{2+} \quad E^{\ominus} = +0.77 \text{ V}$$
$$Cl_2 + 2e^- \rightleftharpoons 2Cl^- \quad E^{\ominus} = +1.36 \text{ V}$$

stronger oxidant
stronger reductant

$$Cl_2 + 2e^- \rightleftharpoons 2Cl^-$$
$$Fe^{3+} + e^- \rightleftharpoons Fe^{2+}$$

REACTANTS: $Cl_2 + Fe^{2+}$
PRODUCTS: $Cl^- + Fe^{3+}$

'WRONG' PREDICTIONS

Sometimes, reactions predicted by this method do not happen. The reason for this is either

- the reaction has a very high activation energy ($E^{\ominus}$ values only tell us about the relative energies of the reactants and products, not about the reaction pathway) or

- the actual conditions of the reaction are not standard and so the actual E values are different from the standard $E^{\ominus}$ values.

Balancing and using redox equations

BALANCING REDOX EQUATIONS

Once you have predicted that a reaction will happen:

1. Copy down the two half equations from the data book with
   - the more positive one on top
   - the more negative one underneath and reversed so that all the reactants are on the left.

2. Scale the two equations so that the number of electrons involved in each is the same.

3. Add the two half equations together.

4. Cancel out any ions or molecules which appear on both sides of the equation.

e.g. For the reaction between manganate (VII) and hydrogen peroxide

MnO_4^-/Mn^{2+} $E^\ominus = +1.58$; O_2/H_2O_2 $E^\ominus = +0.68$

				$E^\ominus$
$MnO_4^- + 8H^+ + 5e^-$	$\rightleftharpoons$	Mn^{2+} $4H_2O$		$+1.58$
H_2O_2	$\rightleftharpoons$	$O_2 + 2H^+ + 2e^-$		$+0.68$

$2MnO_4^- + 16H^+ + 10e^- \rightleftharpoons 2Mn^{2+} + 8H_2O$
$5H_2O_2 \rightleftharpoons 5O_2 + 10H^+ + 10e^-$

$2MnO_4^- + 16H^+ + 5H_2O_2 \rightleftharpoons 2Mn^{2+} + 8H_2O + 5O_2 + 10H^+$

$2MnO_4^- + 6H^+ + 5H_2O_2 \rightleftharpoons 2Mn^{2+} + 8H_2O + 5O_2$

PREDICTING INFORMATION ABOUT CELLS FROM $E^\ominus$ VALUES

$E^\ominus$ values can be used to predict changes which would happen in a cell made up of two half cells.

For example, take a cell made up of a copper half cell and an iron(III)/iron(II) half cell.

Look up the $E^\ominus$ values and write them on a number line like this

this is the negative terminal so will produce electrons

$Cu(s) \Rightarrow Cu^{2+}(aq) + 2e^-$

during the reaction:
copper will dissolve
$[Cu^{2+}]$ will increase

this is the positive terminal so will use up electrons

$Fe^{3+} + e^- \Rightarrow Fe^{2+}$

during the reaction:
$[Fe^{3+}]$ will decrease
$[Fe^{2+}]$ will increase

CHANGING CONDITIONS IN THE CELLS

The reaction in each half cell is an equilibrium whose position can be changed by changing the concentrations of the ions.

So increasing $[Fe^{3+}]$ will increase the cell potential

while increasing $[Cu^{2+}]$ will decrease the cell potential.

Inorganic chemistry: introduction

A good inorganic answer:

> *makes a statement*
> *gives an example or illustrates the statement*
> *explains it using a theory*

It will be a *factual statement* about *physical* or *chemical properties* which is *explained* by *theories* that you have learnt in physical chemistry.

---

e.g. **Facts**

listed with an example

Physical properties such as:
state and fixed points (m.p. or b.p.)
metal or non-metal
conductor or insulator
hard or soft
ductile or brittle

Chemical properties:
reactions with water, acid, or base

reactions with metals: Na, Mg, Fe, Cu
and non-metals: O_2, Cl_2

reactions with ammonia, chloride or
fluoride ions, etc.

behaviour in solvents

Theories

theories and ideas used

explained in terms of
structure and bonding

explained in terms of
acid–base equilibria

explained in terms of
redox equilibria

explained in terms of
complex ion equilibria

explained in terms of
solubility equilibria

The key to all inorganic chemistry is the periodic table. Its patterns help you to learn the facts and to explain them.

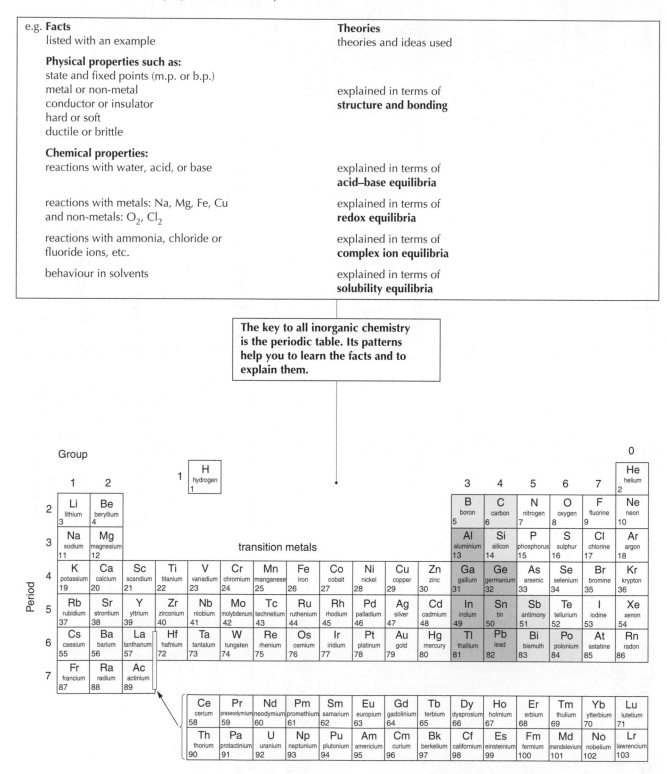

Periodic patterns

PATTERN 1. METAL/NON-METAL TRENDS

Metals

physical	conduct
	ductile/malleable
chemical	basic oxides
	reducing agents
	form cations

Non-metals

physical	insulate
	brittle
chemical	acidic oxides
	oxidising agents
	form anions

PATTERN 2. S, D, AND P BLOCKS

s block

physical	soft, low m.p. metals
	colourless compounds
	alkaline in water
	strong reducing agents
chemical	valency same as group number
	unreactive cations: little hydrolysis

d block

physical	hard, high m.p. metals
	coloured compounds
chemical	unreactive in water
	variable valency
	form complex ions
	cations hydrolysed

p block metals

physical	softer and lower m.p. than d block
chemical	two valencies except Al

p block non-metals

physical	solids, liquids, or gases depending on structure
chemical	variable valency

p

d

s

PATTERN 3. DOWN THE GROUPS

Similar properties within a group related to the outer shell electrons

Trends and differences going down a group related to the number of inner electrons and hence the size of the atoms.

The atoms of the elements at the top of a group are often so small that they have unusual properties.

1 2

3 4 5 6 7 8

increasing number of inner snells so trends

same outer shell so similar properties

PATTERN 4. ACROSS THE PERIODS

physical

metal → non-metal
conduct → insulate
ductile → brittle

chemical

basic in water → acidic in water
reductants → oxidants
cations → anions or ligands

Patterns in the elements and compounds of the second period

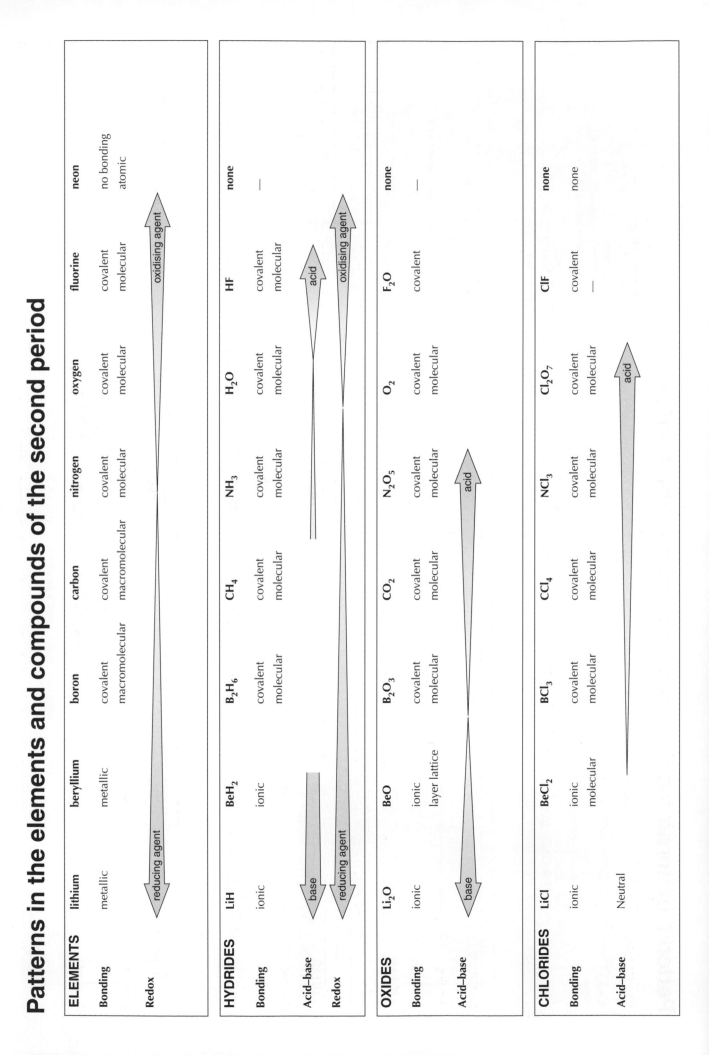

ELEMENTS	lithium	beryllium	boron	carbon	nitrogen	oxygen	fluorine	neon
Bonding	metallic	metallic	covalent macromolecular	covalent macromolecular	covalent molecular	covalent molecular	covalent molecular	no bonding atomic
Redox	← reducing agent ———————————————— oxidising agent →							

HYDRIDES	LiH	BeH₂	B₂H₆	CH₄	NH₃	H₂O	HF	none
Bonding	ionic	ionic	covalent molecular	covalent molecular	covalent molecular	covalent molecular	covalent molecular	—
Acid–base	← base					acid →		
Redox	← reducing agent ———————————————— oxidising agent →							

OXIDES	Li₂O	BeO	B₂O₃	CO₂	N₂O₅	O₂	F₂O	none
Bonding	ionic	ionic layer lattice	covalent molecular	covalent molecular	covalent molecular	covalent molecular	covalent	—
Acid–base	← base ———— acid →				acid →			

CHLORIDES	LiCl	BeCl₂	BCl₃	CCl₄	NCl₃	Cl₂O₇	ClF	none
Bonding	ionic	ionic molecular	covalent molecular	covalent molecular	covalent molecular	covalent molecular	covalent	none
Acid–base	Neutral	← acid					—	none

Patterns in the elements and compounds of the third period

ELEMENTS

	sodium	magnesium	aluminium	silicon	phosphorus	sulphur	chlorine	argon
Bonding	metallic	metallic	metallic	covalent macromolecular	covalent molecular	covalent molecular	covalent molecular	no bonding atomic
Redox	← reducing agent						oxidising agent →	

HYDRIDES

	NaH	MgH$_2$	AlH$_3$	SiH$_4$	PH$_3$	H$_2$S	HCl	none
Bonding	ionic	ionic	covalent	covalent molecular	covalent molecular	covalent molecular	covalent molecular	—
Acid–base	base ←					acid →		
Redox	← reducing agent					oxidising agent →		

OXIDES

	Na$_2$O	MgO	Al$_2$O$_3$	SiO$_2$	P$_4$O$_{10}$	SO$_2$	Cl$_2$O	none
Bonding	ionic	ionic	ionic	covalent macromolecular	covalent molecular	covalent molecular	covalent molecular	—
Acid–base	base ←		amphoteric			acid →		

CHLORIDES

	NaCl	MgCl$_2$	AlCl$_3$	SiCl$_4$	PCl$_5$	SCl$_2$	Cl$_2$	none
Bonding	ionic	ionic	ionic/covalent	covalent molecular	covalent molecular	covalent molecular	covalent molecular	—
Acid–base	neutral					acid →		

Group 1 chemistry: key facts

Elements   all very reactive and so kept under oil

all soft and silvery

Compounds   all *ionic*

all contain the group 1 metal as a cation in the **+I** oxidation state

all are *soluble*

when heated all are *more stable* than the corresponding group 2 compound

Stability

Compounds of group 1 are generally stable to heat and usually simply melt.

Exceptions are the nitrate and hydrogen carbonate:

$$2NaNO_3(s) \xrightarrow{\text{heat}} 2NaNO_2(s) + O_2(g)$$

$$2NaHCO_3(s) \xrightarrow{\text{heat}} Na_2CO_3(s) + H_2O(g) + CO_2(g)$$

Trends down the group
reaction with oxygen

$4Li(s) + O_2(g) \rightarrow 2Li_2O(s)$ OXIDE

$2Na(s) + O_2(g) \rightarrow Na_2O_2(s)$ PEROXIDE

$K(s) + O_2(g) \rightarrow KO_2(s)$ SUPEROXIDE

Anomalous or unusual properties of the first member of the group, lithium

1. Lithium carbonate is unstable:

$$Li_2CO_3(s) \xrightarrow{\text{heat}} Li_2O(s) + CO_2(g)$$

2. Lithium hydroxide is unstable to heat:

$$2Li_2OH(s) \xrightarrow{\text{heat}} Li_2O(s) + H_2O(g)$$

3. Lithium hydroxide is sparingly soluble.

4. Lithium forms some covalent compounds.

Explain these facts in terms of the high polarising power of the lithium ion due to its small size. This means that the positive charge of the ion is concentrated in a small volume giving it high charge density and the ability to attract nearby regions of negative charge.

1
Li
Na
K
Rb
Cs

more reactive

Acid–base reactions
1. With water

All the elements react rapidly with water forming alkaline solutions:

$$2Na(s) + 2H_2O(l) \rightarrow 2Na^+(aq) + 2OH^-(aq) + H_2(g)$$

2. The compounds

The anions of some of the compounds are good bases, so after hydration, they are hydrolysed, reacting with water to make alkaline solutions:

HYDRATION

$Na_2O(s) + H_2O(l) \rightarrow 2Na^+(aq) + O^{2-}(aq)$

$NaH(s) + H_2O(l) \rightarrow Na^+(aq) + H^-(aq)$

$Na_2CO_3(s) + H_2O(l) \rightarrow 2Na^+(aq) + CO_3^{2-}(aq)$

HYDROLYSIS

$O^{2-}(aq) + H_2O(l) \rightarrow 2OH^-(aq)$

$H^-(aq) + H_2O(l) \rightarrow H_2(g) + OH^-(aq)$

$CO_3^{2-}(aq) + H_2O(l) \rightarrow HCO_3^-(aq) + OH^-(aq)$

Precipitation and solubility
All group 1 compounds are soluble.

Solutions of their hydroxides and carbonates can be used to precipitate the hydroxide and carbonates of other metals:

$M^{2+}(aq) + 2OH^-(aq) \rightarrow M(OH)_2(s)$

$M^{2+}(aq) + CO_3^{2-}(aq) \rightarrow MCO_3(s)$

Redox reactions
The metals are strong reducing agents and react with most non-metals producing ionic products

$$2Na(s) + H_2(g) \xrightarrow{\text{heat}} 2NaH(s)$$

$$2Na(s) + Cl_2(g) \xrightarrow{\text{heat}} 2NaCl(s)$$

$$2Na(s) + air \longrightarrow Na_2O(s)$$

The reaction with water is also a redox reaction.

Group 1 chemistry: key ideas

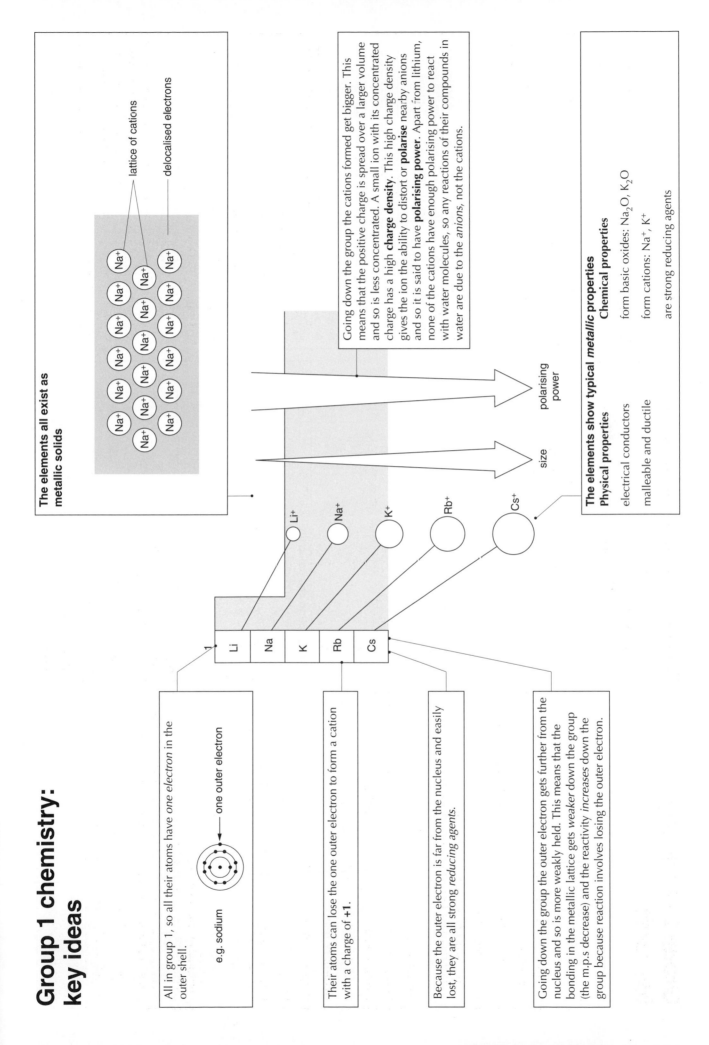

The elements all exist as metallic solids

lattice of cations

delocalised electrons

| Na^+ | Na^+ | Na^+ |
| Na^+ | Na^+ | Na^+ |

Going down the group the cations formed get bigger. This means that the positive charge is spread over a larger volume and so is less concentrated. A small ion with its concentrated charge has a high **charge density**. This high charge density gives the ion the ability to distort or **polarise** nearby anions and so it is said to have **polarising power**. Apart from lithium, none of the cations have enough polarising power to react with water molecules, so any reactions of their compounds in water are due to the *anions*, not the cations.

The elements show typical *metallic* properties

Physical properties	Chemical properties
electrical conductors	form basic oxides: Na_2O, K_2O
malleable and ductile	form cations: Na^+, K^+
	are strong reducing agents

polarising power

size

All in group 1, so all their atoms have *one electron* in the outer shell.

e.g. sodium — one outer electron

Their atoms can lose the one outer electron to form a cation with a charge of **+1.**

Because the outer electron is far from the nucleus and easily lost, they are all strong *reducing agents.*

Going down the group the outer electron gets further from the nucleus and so is more weakly held. This means that the bonding in the metallic lattice gets *weaker* down the group (the m.p.s decrease) and the reactivity *increases* down the group because reaction involves losing the outer electron.

| 1 |
| Li |
| Na |
| K |
| Rb |
| Cs |

Li^+ Na^+ K^+ Rb^+ Cs^+

Group 2 chemistry: key facts

Elements

all quite reactive, but less so than group 1

all silvery, but harder than group 1

all form oxide layers quickly which stops further reaction

Compounds

all *ionic*, but some covalent character seen in beryllium compounds

all contain the group 2 metal as a cation in the **+II** oxidation state

some compounds are *insoluble*: carbonates, hydroxides, some sulphates (see below)

all are *less stable* when heated than the corresponding group 1 compound

Acid–base reactions

1. With water

All the elements react with water forming alkaline suspensions:

$Ca(s) + 2H_2O(l) \rightarrow Ca^{2+}(aq) + 2OH^-(aq) + H_2(g)$

then $Ca^{2+}(aq) + 2OH^-(aq) \rightleftharpoons Ca(OH)_2(s)$

2. The compounds

Like group 1, the anions of some of the compounds are good bases. If they dissolve, after hydration, they are hydrolysed, reacting with water to make alkaline suspensions.

3. With acids

Being bases, the anions of many of the compounds are protonated giving soluble products

$MgO(s) + 2H_3O^+(aq) \rightarrow Mg^{2+}(aq) + 3H_2O(l)$

$CaCO_3(s) + 2H_3O^+(aq) \rightarrow Ca^{2+}(aq) + 3H_2O(l) + CO_2(g)$

Redox reactions

The metals are strong reducing agents and react with most non-metals producing ionic products

The reaction of the metals with water is also a redox reaction.

more reactive

2
Be
Mg
Ca
Sr
Ba

Anomalous or unusual properties of the first member of the group, beryllium

1. Beryllium oxide is amphoteric

as base: $BeO(s) + 2H_3O^+(aq) \rightarrow Be^{2+}(aq) + 3H_2O(l)$

as acid: $BeO(s) + 2OH^-(aq) + H_2O(l) \rightarrow Be(OH)_4^-(aq)$

2. Beryllium chloride forms a layer lattice rather than an ionic one. In this way it is like aluminium chloride. Beryllium and aluminium are diagonal neighbours in the periodic table and this is an example of what is sometimes called a **diagonal relationship**.

Explain these facts in terms of the high polarising power of the beryllium ion due to its small size, which means that its compounds show some covalent character and anions near the cation are distorted.

Stability

Group 2 compounds are generally less stable than the equivalent group 1 compound because of the greater polarising power of the cation. This can distort the anion. Because the size of the cations changes down the group there are trends in stability.

Stability of the carbonates, nitrates, and hydroxides

$MgCO_3(s) \xrightarrow{\text{heat}} MgO(s) + CO_2(g)$

and $2Ca(NO_3)_2(s) \xrightarrow{\text{heat}} 2CaO(s) + 4NO_2(g) + O_2(g)$

stability increases down the group

e.g.
$MgCO_3$	decomposes at 540 °C
$CaCO_3$	decomposes at 900 °C
$SrCO_3$	decomposes at 1290 °C
$BaCO_3$	decomposes at 1360 °C

Smaller ions have greater polarising power.

Precipitation and solubility

Some of the compounds are insoluble or slightly soluble and may redissolve if they react, e.g. lime water and its reaction with carbon dioxide.

$Ca^{2+}(aq) + 2OH^-(aq) + CO_2(g) \rightarrow CaCO_3(s) + H_2O(l)$

then $CaCO_3(s) + H_2O(l) + CO_2(g) \rightarrow Ca^{2+}(aq) + 2HCO_3^-(aq)$

There are trends in solubility down the group. These change from one kind of compound to the next

Solubility of the hydroxides and sulphates:
hydroxide solubility **increases**, but **sulphate** solubility **decreases** down the group

$Mg(OH)_2$	insoluble	$MgSO_4$	soluble
$Ca(OH)_2$	slightly soluble	$CaSO_4$	slightly soluble
$Ba(OH)_2$	quite soluble	$BaSO_4$	insoluble

Explain using an enthalpy cycle: trend is dominated either by lattice or hydration energy.

Group 2 chemistry: key ideas

The elements all exist as metallic solids

The *greater charge* and *smaller size* of these ions compared to group 1 and the fact that there are *twice as many* delocalised outer electrons accounts for the *greater hardness* and *higher melting points* compared to group 1.

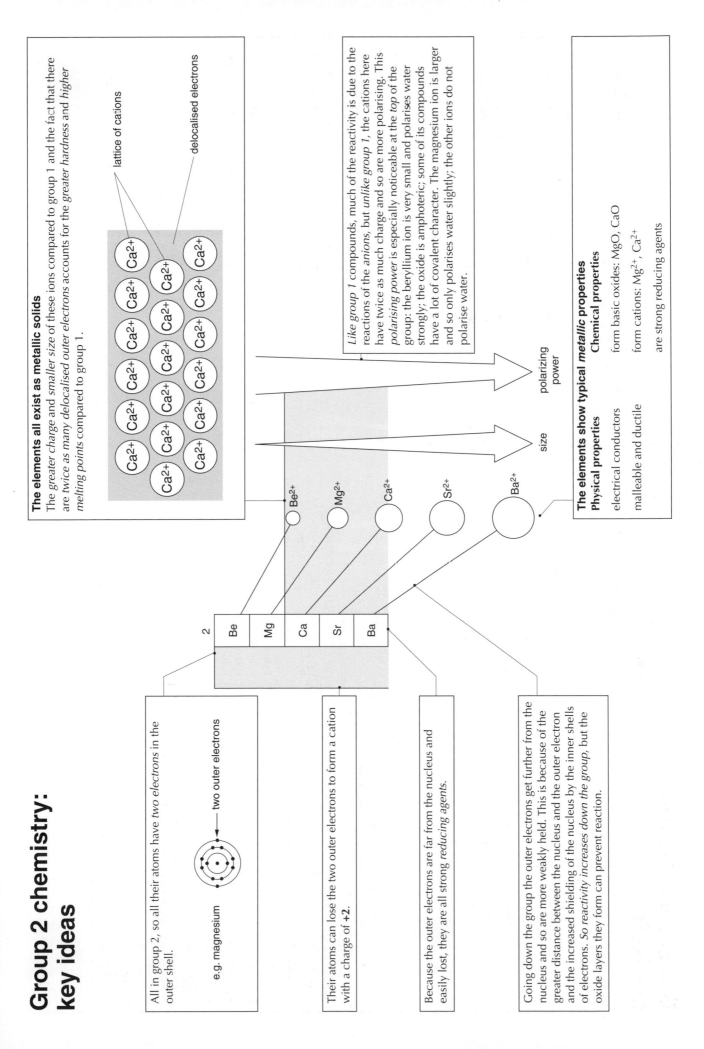

lattice of cations

delocalised electrons

Ca^{2+}

Like group 1 compounds, much of the reactivity is due to the reactions of the *anions*, but *unlike group 1*, the cations here have twice as much charge and so are more polarising. This *polarising power* is especially noticeable at the *top of the* group: the beryllium ion is very small and polarises water strongly; the oxide is amphoteric; some of its compounds have a lot of covalent character. The magnesium ion is larger and so only polarises water slightly; the other ions do not polarise water.

polarizing power

size

Be^{2+}

Mg^{2+}

Ca^{2+}

Sr^{2+}

Ba^{2+}

The elements show typical *metallic properties*

Physical properties

electrical conductors

malleable and ductile

Chemical properties

form basic oxides: MgO, CaO

form cations: Mg^{2+}, Ca^{2+}

are strong reducing agents

2
Be
Mg
Ca
Sr
Ba

All in group 2, so all their atoms have *two electrons* in the outer shell.

two outer electrons

e.g. magnesium

Their atoms can lose the two outer electrons to form a cation with a charge of **+2**.

Because the outer electrons are far from the nucleus and easily lost, they are all strong *reducing agents*.

Going down the group the outer electrons get further from the nucleus and so are more weakly held. This is because of the greater distance between the nucleus and the outer electron and the increased shielding of the nucleus by the inner shells of electrons. *So reactivity increases down the group,* but the oxide layers they form can prevent reaction.

Aluminium chemistry: key facts

Properties related to the periodic table

The properties of the element and its compounds are related to the position of the element in the periodic table. It is near the borderline between metals and non-metals.

The metal is a strong reducing agent, reducing hydrogen in both acid and alkaline conditions.

It is used in thermite reactions in which aluminium powder reduces metal oxides.

Physical properties of the element

Aluminium has a huge range of uses due to its high strength/weight ratio, resistance to corrosion (the result of its protective oxide layer), and its good conductivity. It is one of the two most important industrial metals.

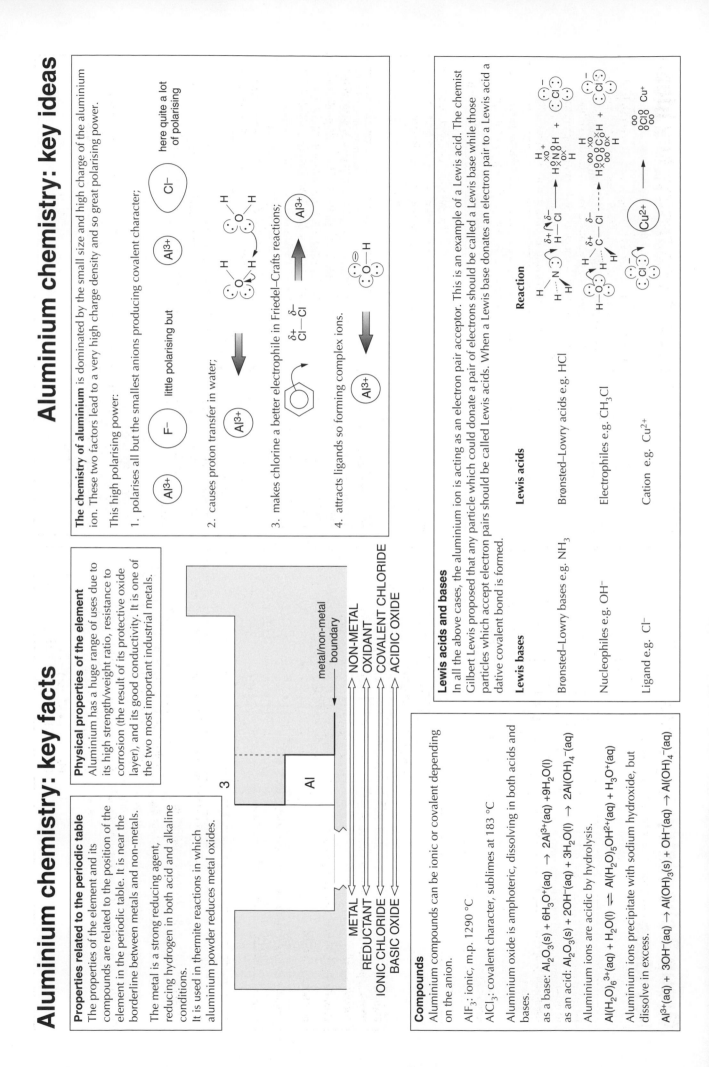

3

Al

metal/non-metal
↓ boundary

METAL → NON-METAL
REDUCTANT → OXIDANT
IONIC CHLORIDE → COVALENT CHLORIDE
BASIC OXIDE → ACIDIC OXIDE

Compounds

Aluminium compounds can be ionic or covalent depending on the anion.

AlF_3: ionic, m.p. 1290 °C

$AlCl_3$: covalent character, sublimes at 183 °C

Aluminium oxide is amphoteric, dissolving in both acids and bases.

as a base: $Al_2O_3(s) + 6H_3O^+(aq) \rightarrow 2Al^{3+}(aq) + 9H_2O(l)$

as an acid: $Al_2O_3(s) + 2OH^-(aq) + 3H_2O(l) \rightarrow 2Al(OH)_4^-(aq)$

Aluminium ions are acidic by hydrolysis.

$Al(H_2O)_6^{3+}(aq) + H_2O(l) \rightleftharpoons Al(H_2O)_5OH^{2+}(aq) + H_3O^+(aq)$

Aluminium ions precipitate with sodium hydroxide, but dissolve in excess.

$Al^{3+}(aq) + 3OH^-(aq) \rightarrow Al(OH)_3(s) + OH^-(aq) \rightarrow Al(OH)_4^-(aq)$

Aluminium chemistry: key ideas

The chemistry of aluminium is dominated by the small size and high charge of the aluminium ion. These two factors lead to a very high charge density and so great polarising power.

This high polarising power:

1. polarises all but the smallest anions producing covalent character;

Al^{3+} F^- little polarising but

Al^{3+} Cl^- here quite a lot of polarising

2. causes proton transfer in water;

Al^{3+}

3. makes chlorine a better electrophile in Friedel–Crafts reactions;

$\overset{\delta+}{Cl}—\overset{\delta-}{Cl}$

Al^{3+}

4. attracts ligands so forming complex ions.

Al^{3+}

Lewis acids and bases

In all the above cases, the aluminium ion is acting as an electron pair acceptor. This is an example of a Lewis acid. The chemist Gilbert Lewis proposed that any particle which could donate a pair of electrons should be called Lewis bases while those particles which accept electron pairs should be called Lewis acids. When a Lewis base donates an electron pair to a Lewis acid a dative covalent bond is formed.

Lewis bases	Lewis acids	Reaction
Brønsted–Lowry bases e.g. NH_3	Brønsted–Lowry acids e.g. HCl	
Nucleophiles e.g. OH^-	Electrophiles e.g. CH_3Cl	
Ligand e.g. Cl^-	Cation e.g. Cu^{2+}	Cu^+

Group 4 chemistry: key ideas

The increase in atomic size as the group is descended leads to a loss of control over the outer electrons. Instead of electrons being localised and held tightly in covalent bonds as they are at the top, they are delocalised and free to move in the metals. This trend is not clear cut, because graphite has delocalised electrons and conducts like a metal while below 13 °C tin is most stable as a macromolecular grey solid.

The inert pair effect: The outer shell for group 4 is s^2p^2 — four outer electrons available for bonding and so an oxidation state of +IV. But as the group is descended the +IV oxidation state becomes less stable with respect to the +II oxidation state. This trend is often called the *inert pair effect* because it appears that the two s electrons have become *inert* and less available for bonding leaving only the two electrons in the p orbitals able to take part in bonding.

Hybridisation and p–π overlap

The ground state for carbon is $1s^2 2s^2 2p^2$, which means that the outer shell electrons are of two kinds, s and p, and yet carbon forms four identical bonds. The ground state is only the lowest energy state; an alternative state of higher energy has the four outer electrons in sp^3 hybrid orbitals which have some s and some p character. These s–p hybrid orbitals can overlap better than s or p orbitals so producing stronger and identical bonds.

Elements in the second period of the periodic table have p orbitals of the same size as oxygen and so form strong pi bonds with oxygen. Silicon in the third period has larger p orbitals which do not match in size and overlap well with oxygen's p orbitals so strong pi bonds are not formed.

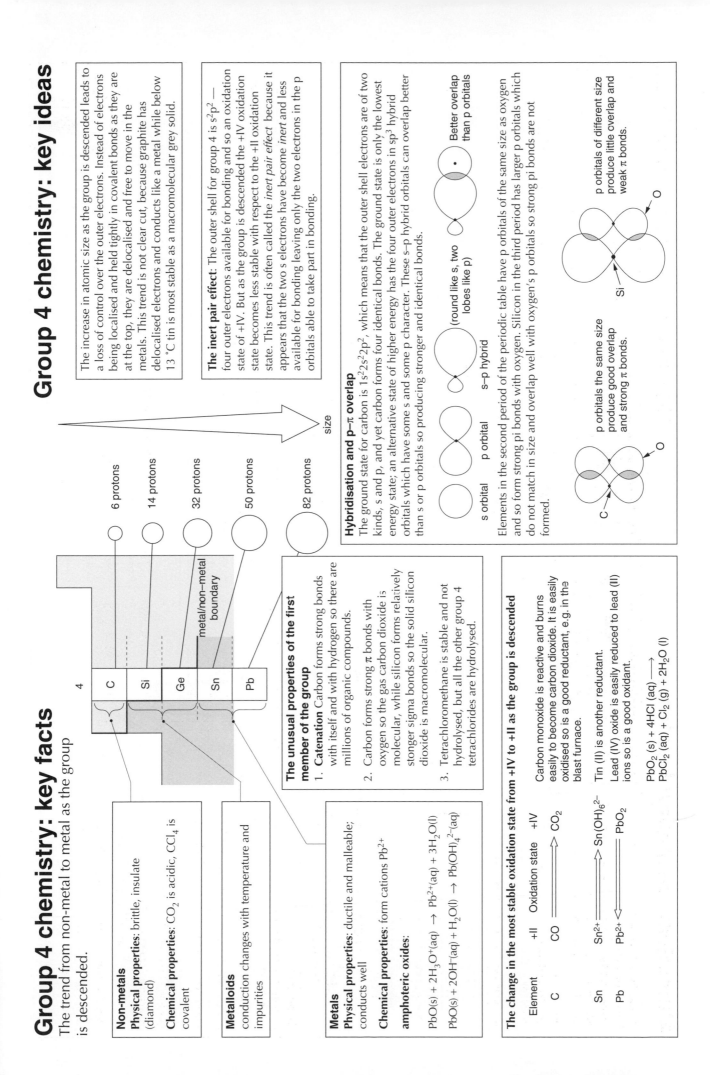

size

6 protons

14 protons

32 protons

50 protons

82 protons

metal/non–metal boundary

4

C

Si

Ge

Sn

Pb

Non-metals

Physical properties: brittle, insulate (diamond)

Chemical properties: CO_2 is acidic, CCl_4 is covalent

Metalloids

conduction changes with temperature and impurities

Metals

Physical properties: ductile and malleable; conducts well

Chemical properties: form cations Pb^{2+}

amphoteric oxides:

$PbO(s) + 2H_3O^+(aq) \rightarrow Pb^{2+}(aq) + 3H_2O(l)$

$PbO(s) + 2OH^-(aq) + H_2O(l) \rightarrow Pb(OH)_4^{2-}(aq)$

round like s, two lobes like p

Better overlap than p orbitals

s orbital   p orbital   s–p hybrid

p orbitals the same size produce good overlap and strong π bonds.

p orbitals of different size produce little overlap and weak π bonds.

O

Si

O

C

Group 4 chemistry: key facts

The trend from non-metal to metal as the group is descended.

The unusual properties of the first member of the group

1. **Catenation** Carbon forms strong bonds with itself and with hydrogen so there are millions of organic compounds.

2. Carbon forms strong π bonds with oxygen so the gas carbon dioxide is molecular, while silicon forms relatively stonger sigma bonds so the solid silicon dioxide is macromolecular.

3. Tetrachloromethane is stable and not hydrolysed, but all the other group 4 tetrachlorides are hydrolysed.

The change in the most stable oxidation state from +IV to +II as the group is descended

Element	+II	Oxidation state	+IV	
C		$CO \xrightarrow{\hspace{1cm}} CO_2$		Carbon monoxide is reactive and burns easily to become carbon dioxide. It is easily oxidised so is a good reductant, e.g. in the blast furnace.
Sn		$Sn^{2+} \xrightleftharpoons{\hspace{1cm}} Sn(OH)_6^{2-}$		Tin (II) is another reductant.
Pb		$Pb^{2+} \xleftarrow{\hspace{1cm}} PbO_2$		Lead (IV) oxide is easily reduced to lead (II) ions so is a good oxidant.

$PbO_2 (s) + 4HCl (aq) \longrightarrow$
$PbCl_2 (aq) + Cl_2 (g) + 2H_2O (l)$

Nitrogen chemistry: key facts

AMMONIA

Properties due to the lone pair

1. Base

$$NH_3(g) + HCl(g) \rightarrow NH_4Cl(s) \quad \text{and} \quad NH_3(g) + H_2O(l) \rightleftharpoons NH_4^+(aq) + OH^-(aq)$$

2. Nucleophile

$$2NH_3(g) + CH_3COCl(l) \rightarrow CH_3CONH_2(s) + NH_4Cl(s)$$

3. Ligand

$$4NH_3(aq) + Cu^{2+}(aq) \rightarrow Cu(NH_3)_4{}^{2+}(aq)$$

Redox properties

Nitrogen in ammonia has an oxidation state of –III and so can be oxidised to one of the many higher oxidation states of nitrogen. Ammonia is therefore a reducing agent.

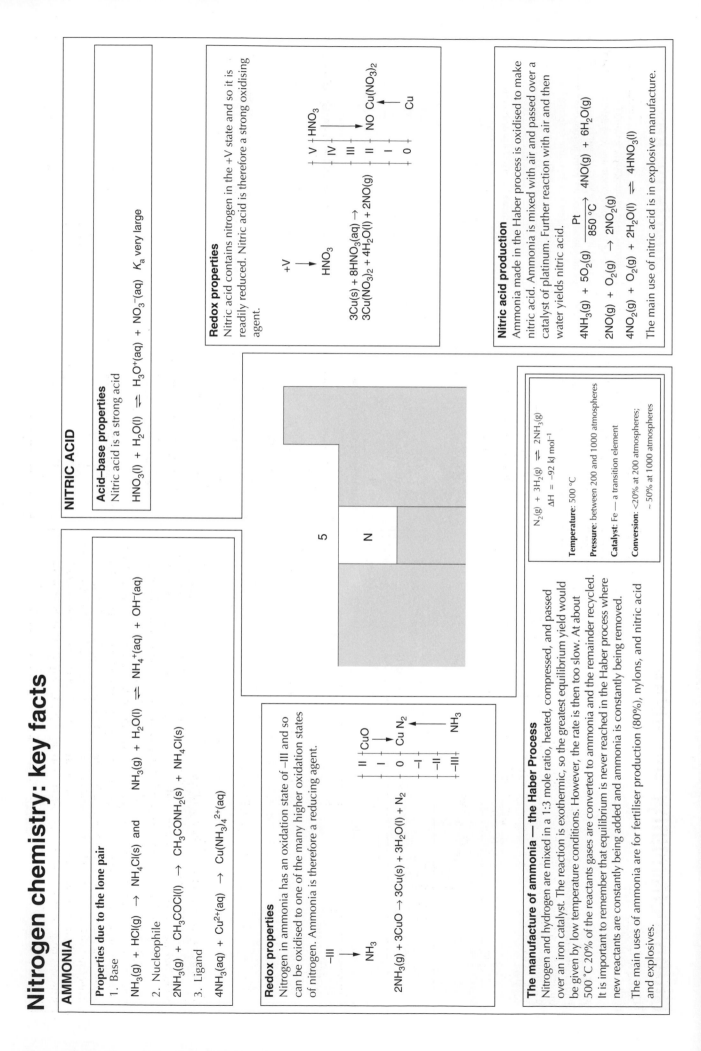

$$2NH_3(g) + 3CuO \rightarrow 3Cu(s) + 3H_2O(l) + N_2$$

The manufacture of ammonia — the Haber Process

Nitrogen and hydrogen are mixed in a 1:3 mole ratio, heated, compressed, and passed over an iron catalyst. The reaction is exothermic, so the greatest equilibrium yield would be given by low temperature conditions. However, the rate is then too slow. At about 500 °C 20% of the reactants gases are converted to ammonia and the remainder recycled. It is important to remember that equilibrium is never reached in the Haber process where new reactants are constantly being added and ammonia is constantly being removed.

The main uses of ammonia are for fertiliser production (80%), nylons, and nitric acid and explosives.

$$N_2(g) + 3H_2(g) \rightleftharpoons 2NH_3(g)$$
$$\Delta H = -92 \text{ kJ mol}^{-1}$$

Temperature: 500 °C

Pressure: between 200 and 1000 atmospheres

Catalyst: Fe — a transition element

Conversion: <20% at 200 atmospheres; ~ 50% at 1000 atmospheres

NITRIC ACID

Acid–base properties

Nitric acid is a strong acid

$$HNO_3(l) + H_2O(l) \rightleftharpoons H_3O^+(aq) + NO_3^-(aq) \quad K_a \text{ very large}$$

Redox properties

Nitric acid contains nitrogen in the +V state and so it is readily reduced. Nitric acid is therefore a strong oxidising agent.

$$3Cu(s) + 8HNO_3(aq) \rightarrow 3Cu(NO_3)_2 + 4H_2O(l) + 2NO(g)$$

Nitric acid production

Ammonia made in the Haber process is oxidised to make nitric acid. Ammonia is mixed with air and passed over a catalyst of platinum. Further reaction with air and then water yields nitric acid.

$$4NH_3(g) + 5O_2(g) \xrightarrow[850\,°C]{Pt} 4NO(g) + 6H_2O(g)$$

$$2NO(g) + O_2(g) \rightarrow 2NO_2(g)$$

$$4NO_2(g) + O_2(g) + 2H_2O(l) \rightleftharpoons 4HNO_3(l)$$

The main use of nitric acid is in explosive manufacture.

Nitrogen chemistry: key ideas

Nitrogen is an element

Nitrogen is a covalently bonded, molecular gas. It exists as diatomic molecules. The triple bond between the nitrogen atoms is very strong so nitrogen is very unreactive.

Phosphorus in the same group exists as P_4 molecules (white phosphorus) with single bonds between the atoms. These bonds are weak and there is bond strain, because the bond angles are only 60°. This makes phosphorus very reactive.

944 kJ mol^{-1}

$: N \equiv N :$

172 kJ mol^{-1}

bond angles of 60°

Nitrogen is a non-metal

It is in period 2, so its atoms have two shells of electrons around the nucleus. This means that it cannot expand its outer shell, but can only have a maximum of eight electrons around the nucleus.

Nitrogen is in group 5, so there are five electrons in the outer shell.

5

N

This means a nitrogen atom can:

1. Gain three electrons becoming a nitride ion when it reacts with very reactive metals.

   The nitride ion is a strong base.

   N^{3-} $3\ Mg(s) + N_2(g) \longrightarrow Mg_3N_2(s)$

   $N^{3-}(s) + 3\ H_2O(l) \longrightarrow NH_3(aq) + 3\ OH^-(aq)$

2. Share three electrons making three covalent bonds.

   The remaining lone pair on the nitrogen dominates the properties of the resulting molecule, allowing it to behave as:

   (i) a base when it attacks slightly positive hydrogen atoms

   $\delta+\ \ \delta-$
   $H — Cl$ ⟶ ACID

   (ii) a nucleophile when it attacks slightly positive non-metal atoms

   $\delta+\ \ \delta-$
   $H — C — Cl$ ⟶ ELECTROPHILE

   (iii) a ligand when it attacks positive metal ions

   Cu^{2+} ⟶ CATION

3. Form three bonds by sharing and a fourth using the lone pair in a dative covalent bond.

The importance of nitrogen in living things

All living things contain nitrogen in the form of proteins (polymers of amino-acids) and nucleic acids. The main source of the nitrogen needed to make these compounds is the atmosphere. But the nitrogen in the atmosphere, while being very abundant, is very unreactive because of the strong nitrogen–nitrogen triple bond. The process of converting this unreactive atmospheric nitrogen into some reactive form which plants can take up is called fixation. The natural forms of fixation include the lightning in thunderstorms and nitrogen-fixing bacteria of the roots of leguminous plants. These natural processes have not met the increasing demands of growing human populations. Man-made methods of fixation have been introduced of which the Haber process is the best known. The excessive use of fertilisers results in the leaching of unused fertiliser into rivers causing rapid algal growth. This depletes the oxygen in the water which cannot then support aquatic life.

Oxygen chemistry: key facts

Compounds

Oxygen forms both ionic compounds with metals and covalent compounds with non-metals.

$$2Ca(s) + O_2(g) \rightarrow 2CaO(s)$$

$$S(s) + O_2(g) \rightarrow SO_2(g)$$

Many elements form oxo-compounds in which the element is oxidised by the oxygen into a high oxidation state.

+IV	+VI	+V	+VII	+VI
↓	↓	↓	↓	↓
$CaCO_3$	$MgSO_4$	KNO_3	$KMnO_4$	$K_2Cr_2O_7$

Peroxides are compounds containing an oxygen–oxygen covalent, sigma bond. This is quite a weak bond and so peroxides are unstable. Hydrogen peroxide disproportionates into oxygen and water.

$$146 \text{ kJ mol}^{-1}$$

$$2H_2O_2(l) \longrightarrow 2H_2O(l) + O_2(g)$$

0	—	O_2
–I	—	H_2O_2
–II	—	H_2O

Acid–base reactions

In a sample of water, a very small number of the molecules react with each other forming ions, so pure water conducts very slightly. Equal numbers of hydroxonium and hydroxide ions are formed. Any particle which reacts with water (hydrolysis) upsetting this balance will change the pH of water. Oxide ions do this because they act as bases.

$$H_2O(l) + H_2O(l) \rightleftharpoons H_3O^+(aq) + OH^-(aq) \quad K_w = 1 \times 10^{-14} \text{ mol}^2 \text{ dm}^{-6}$$

$$HSO_4^-(aq) + H_2O(l) \rightarrow H_3O^+(aq) + SO_4^{2-}(aq) \quad [H_3O^+] \text{ increased, solution acidic}$$

$$O^{2-}(s) + 2H_3O^+(aq) \rightarrow 3H_2O(l) \quad [H_3O^+] \text{ reduced, solution alkaline}$$

Redox reactions

Water can act as an oxidising agent with a reactive metal.

$$Ca(s) + 2H_2O(l) \longrightarrow Ca^{2+}(aq) + 2OH^-(aq)$$

II	—	Ca^{2+}	
I	—		H_2O
0	—	Ca	H_2

OZONE

High in the upper atmosphere oxygen molecules are broken down by radiation from the sun.

The atoms formed react with other oxygen molecules making a layer of ozone.

This ozone absorbs more radiation, which would otherwise damage living cells on Earth.

$$O_2(g) \xrightarrow{h\nu} 2O(g)$$

$$O(g) + O_2(g) \longrightarrow O_3(g)$$

$$O_3(g) \xrightarrow{h\nu} O_2(g) + O(g)$$

Usually the ozone is made and decomposed at the same rate so the total amount in the atmosphere remains constant. However, the amount has been decreasing recently, leaving the so-called holes in the ozone layer, because of our use of chlorofluorocarbons as aerosol propellants and refrigerants. These compounds leak into the atmosphere and circulate. When they reach the upper atmosphere they are decomposed by radiation and produce free radicals which react with ozone decreasing the amount in the upper atmosphere.

$$CCl_2F_2(g) \xrightarrow{h\nu} CClF_2^\bullet(g) + Cl^\bullet(g)$$

$$Cl^\bullet(g) + O_3(g) \longrightarrow ClO(g) + O_2(g) \quad \text{ozone removed}$$

Nitrogen oxides made by aircraft engines also react with ozone, further depleting it.

$$NO(g) + O_3(g) \longrightarrow NO_2(g) + O_2(g) \quad \text{ozone removed}$$

The decrease in ozone in the upper atmosphere allows more high energy radiation to pass through the atmosphere causing more biological damage on earth — skin cancer, cataracts, damage to plants.

Oxygen chemistry: key ideas

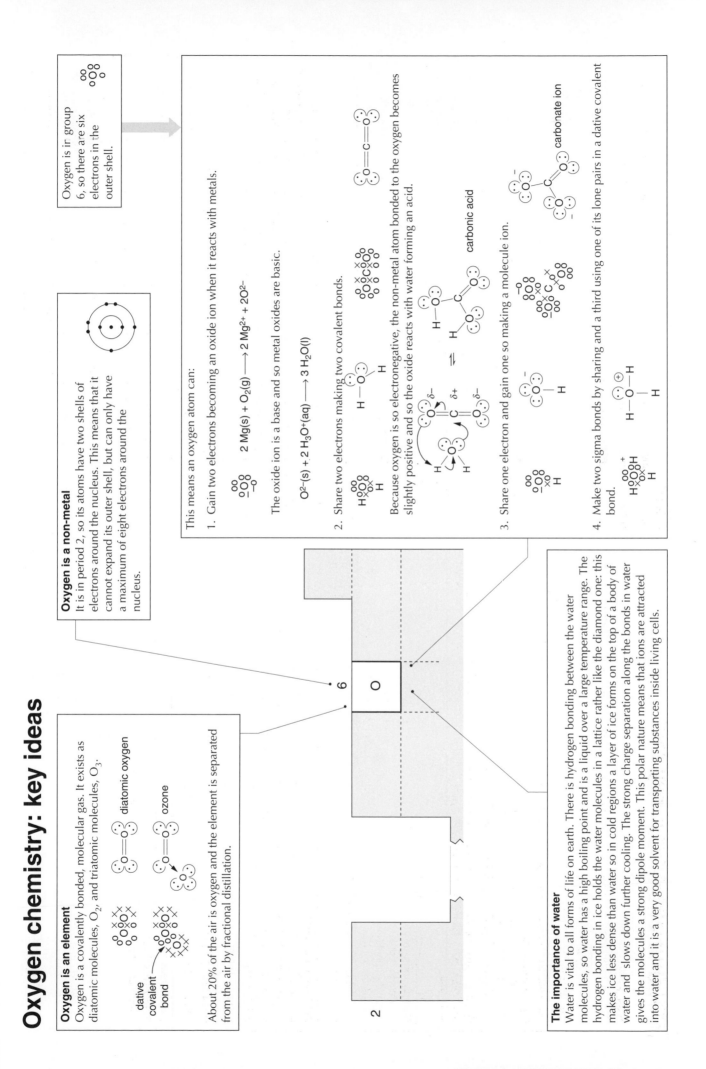

Oxygen is a non-metal

It is in period 2, so its atoms have two shells of electrons around the nucleus. This means that it cannot expand its outer shell, but can only have a maximum of eight electrons around the nucleus.

Oxygen is ir group 6, so there are six electrons in the outer shell.

This means an oxygen atom can:

1. Gain two electrons becoming an oxide ion when it reacts with metals.

$$2\ Mg(s) + O_2(g) \longrightarrow 2\ Mg^{2+} + 2O^{2-}$$

The oxide ion is a base and so metal oxides are basic.

$$O^{2-}(s) + 2\ H_3O^+(aq) \longrightarrow 3\ H_2O(l)$$

2. Share two electrons making two covalent bonds.

Because oxygen is so electronegative, the non-metal atom bonded to the oxygen becomes slightly positive and so the oxide reacts with water forming an acid.

carbonic acid

3. Share one electron and gain one so making a molecule ion.

carbonate ion

4. Make two sigma bonds by sharing and a third using one of its lone pairs in a dative covalent bond.

Oxygen is an element

Oxygen is a covalently bonded, molecular gas. It exists as diatomic molecules, O_2, and triatomic molecules, O_3.

diatomic oxygen

ozone

dative covalent bond

About 20% of the air is oxygen and the element is separated from the air by fractional distillation.

The importance of water

Water is vital to all forms of life on earth. There is hydrogen bonding between the water molecules, so water has a high boiling point and is a liquid over a large temperature range. The hydrogen bonding in ice holds the water molecules in a lattice rather like the diamond one: this makes ice less dense than water so in cold regions a layer of ice forms on the top of a body of water and slows down further cooling. The strong charge separation along the bonds in water gives the molecules a strong dipole moment. This polar nature means that ions are attracted into water and it is a very good solvent for transporting substances inside living cells.

Sulphur chemistry: key facts

Acid–base reactions

hydrogen sulphide is a weak acid

$H_2S(g) + H_2O(l) \rightleftharpoons H_3O^+(aq) + HS^-(aq)$ $K_a = 9 \times 10^{-8}$ mol dm^{-3}

sulphur (IV) oxide — sulphur dioxide — is acidic

$SO_2(g) + H_2O(l) \rightleftharpoons H_2SO_3(aq)$ then $H_2SO_3(aq) + H_2O(l) \rightleftharpoons H_3O^+(aq) + HSO_3^-(aq)$

sulphur (VI) oxide — sulphur trioxide — is acidic

$SO_3(g) + H_2O(l) \rightleftharpoons H_2SO_4(aq)$ then $H_2SO_4(aq) + H_2O(l) \rightleftharpoons H_3O^+(aq) + HSO_4^-(aq)$

sulphuric (VI) acid is a strong dibasic acid — it has two protons

$H_2SO_4(l) + H_2O(l) \rightleftharpoons H_3O^+(aq) + HSO_4^-(aq)$ K_a = very large; $HSO_4^-(aq) + H_2O(l) \rightleftharpoons H_3O^+(aq) + SO_4^{2-}(aq)$ $K_a = 1.2 \times 10^{-2}$ mol dm^{-3}

Compounds

are all covalent except for the metallic sulphides which contain the S^{2-} ion.

Precipitation and solubility

All sulphates except those of barium, calcium, and lead are soluble. So the test for a sulphate is to add barium ions and look for a white precipitate.

Redox reactions

1. sulphur (IV) oxide can be oxidised to sulphur in the (VI) state and so is a good reducing agent

$Cr_2O_7^{2-}(aq) + 3SO_3^{2-}(aq) + 8H_3O^+(aq)$

$\rightarrow 2Cr^{3+}(aq) + 3SO_4^{2-}(aq) + 12H_2O(l)$

2. concentrated sulphuric acid is a strong oxidising agent: e.g. it can oxidise carbon

$2H_2SO_4(l) + C(s) \rightarrow 2H_2O(l) + 2SO_2(g) + CO_2(g)$

PRODUCTION OF SULPHURIC ACID

Sulphuric acid is used in so many industrial processes that is is often used as a primary economic indicator. In other words the more sulphuric acid a country makes, the greater is its economic activity. The UK production is more than 4 million tons a year.

The process:

Sulphur is found 'native' – as the element. It is extracted, melted, and burnt in a furnace:

$$S(l) + O_2(g) \rightarrow SO_2(g)$$

The sulphur dioxide is mixed with air and passed over a heated catalyst of vanadium (V) oxide. The reaction between the air and sulphur dioxide is exothermic. To prevent the catalyst decomposing as the result of the heat of the reaction and the equilibrium position moving to the left, the reaction is cooled when only about 60% is converted and then the mixture passed again over the catalyst. After three passes over the catalyst, almost all the sulphur dioxide is converted into sulphur trioxide.

$2 SO_2(g) + O_2(g) \rightleftharpoons 2 SO_3(g)$ ΔH = −98 kJ mol^{-1}

Conditions: 450°C
2–3 atmospheres
V_2O_5 catalyst

The sulphur trioxide is then hydrolysed (reacted with water) in the presence of sulphuric acid, producing more sulphuric acid.

$SO_3(g) + H_2SO_4(l) \rightarrow H_2S_2O_7(l)$

then $H_2S_2O_7(l) + H_2O(l) \rightarrow 2H_2SO_4(l)$

Uses of sulphuric acid

There are many uses, but the most important ones are in the manufacture of:

1. fertilisers
2. detergents
3. paints and pigments
4. fibres and plastics

Sulphur chemistry: key ideas

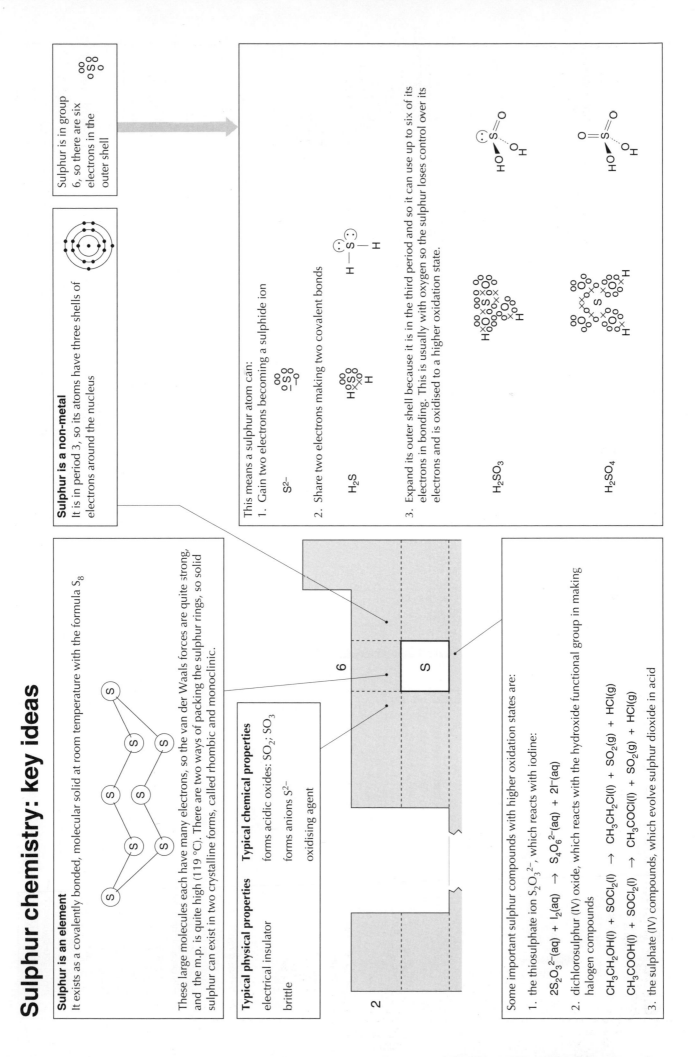

Sulphur is an element

It exists as a covalently bonded, molecular solid at room temperature with the formula S_8

These large molecules each have many electrons, so the van der Waals forces are quite strong, and the m.p. is quite high (119 °C). There are two ways of packing the sulphur rings, so solid sulphur can exist in two crystalline forms, called rhombic and monoclinic.

Typical physical properties   **Typical chemical properties**

electrical insulator   forms acidic oxides: SO_2; SO_3

brittle   forms anions S^{2-}

                                    oxidising agent

Sulphur is a non-metal

It is in period 3, so its atoms have three shells of electrons around the nucleus

Sulphur is in group 6, so there are six electrons in the outer shell

This means a sulphur atom can:
1. Gain two electrons becoming a sulphide ion

   S^{2-}

2. Share two electrons making two covalent bonds

   H_2S

3. Expand its outer shell because it is in the third period and so it can use up to six of its electrons in bonding. This is usually with oxygen so the sulphur loses control over its electrons and is oxidised to a higher oxidation state.

   H_2SO_3

   H_2SO_4

Some important sulphur compounds with higher oxidation states are:

1. the thiosulphate ion $S_2O_3^{2-}$, which reacts with iodine:

   $2S_2O_3^{2-}(aq) + I_2(aq) \rightarrow S_4O_6^{2-}(aq) + 2I^-(aq)$

2. dichlorosulphur (IV) oxide, which reacts with the hydroxide functional group in making halogen compounds

   $CH_3CH_2OH(l) + SOCl_2(l) \rightarrow CH_3CH_2Cl(l) + SO_2(g) + HCl(g)$

   $CH_3COOH(l) + SOCl_2(l) \rightarrow CH_3COCl(l) + SO_2(g) + HCl(g)$

3. the sulphate (IV) compounds, which evolve sulphur dioxide in acid

Group 7 chemistry: key facts

Redox reactions

All the halogens are oxidising agents; the oxidising strength decreases down the group:

$$F_2(g) + 2e^- \rightleftharpoons 2F^-(aq) \qquad E^\ominus = 2.87 \text{ V}$$

$$Cl_2(g) + 2e^- \rightleftharpoons 2Cl^-(aq) \qquad E^\ominus = 1.36 \text{ V}$$

reducing strength

$$Br_2(l) + 2e^- \rightleftharpoons 2Br^-(aq) \qquad E^\ominus = 1.06 \text{ V}$$

oxidising strength

$$I_2(s) + 2e^- \rightleftharpoons 2I^-(aq) \qquad E^\ominus = 0.54 \text{ V}$$

This makes the halide ions increasingly good reducing agents going down the group as the reactions with concentrated sulphuric acid show:

$$Cl^-(s) + H_2SO_4(l) \rightarrow HCl(g) + HSO_4^-(l) \text{ protonation only}$$
$$Br^-(s) + H_2SO_4(l) \rightarrow HBr(g) + HSO_4^-(l) \text{ then HBr oxidised to } Br_2$$
$$I^-(s) + H_2SO_4(l) \rightarrow HI(g) + HSO_4^-(l) \text{ then HI oxidised to } I_2$$

Acid–base reactions

1. reactions of the halogens with water
   apart from fluorine which oxidises water, the halogens all disproportionate in water
   $$Cl_2(g) + 2H_2O(l) \rightleftharpoons H_3O^+(aq) + Cl^-(aq) + HOCl(aq)$$
2. reactions of the halogens with alkali
   The halogens disproportionate in alkali: the products depend on the conditions:
   in *cold, dilute alkali*
   $$Cl_2(g) + 2OH^-(aq) \rightarrow Cl^-(aq) + OCl^-(aq) + H_2O(l)$$
   in *hot, more concentrated alkali*
   $$3Cl_2(g) + 6OH^-(aq) \rightarrow 5Cl^-(aq) + ClO_3^-(aq) + 3H_2O(l)$$
3. reactions of the hydrogen halides with water
   The hydrogen halides are all very soluble in water and form strong acids; acid strength increases down the group due to the decreasing strength of the hydrogen–halogen bond
   $$HHal(g) + H_2O(l) \rightleftharpoons H_3O^+(aq) + Hal^-(aq) \qquad K_a(HI) > K_a(HBr) > K_a(HCl) > K_a(HF)$$

7
F
Cl
Br
I

more reactive

USES OF HALOGENS AND THEIR COMPOUNDS

Chlorine: as a bleach and in the treatment of water; both uses depend on the oxidising power of the chlorine.

Fluorine compounds have three important uses:

- Fluoride ion in toothpastes; F⁻ ion replaces basic OH⁻ ion in tooth enamel so no attack by mouth acids.
- CFCs are compounds of chlorine, fluorine, and carbon; used as aerosol propellants and refrigerants, but cause depletion of ozone layer; replacing chlorine with more fluorine atoms (more strongly bonded to carbon) makes more stable compounds, so CFCs to be replaced by HFAs — **hydrofluoroalkanes** — e.g. CF_3CH_2F.
- Non-stick coatings used in cooking and many other areas are based on polymerised tetrafluoroethene — PTFE.

Chlorine compounds have many uses:

- Dry cleaning solvents are nearly always based on organic chlorine compounds such as 1,1,1-trichloroethane.
- Chloroform is one of the oldest anaesthetics; many others are also based on organic chlorine compounds.
- Herbicides, often known by initials, e.g. 2,4-D, are chlorine compounds which can kill broad-leaved plants.
- Insecticides, DDT is the most famous, are chlorine compounds; they tend to be fat soluble and therefore become concentrated in body tissue as they move up the food chain.
- The halides of silver darken in light; this is a photochemical reaction and is the basis of photographic film.

Elements

- all are reactive; reactivity decreases down the group
- all are coloured; colour darkens down the group
- exist in a range of states at room temperature: F_2 and Cl_2 are gases; Br_2 is liquid; I_2 is solid

Compounds both ionic: NaCl; KI

and covalent: HI; CCl_4; HOBr

apart from fluorine (always −I state) they are found in a range of oxidation states

Complexing

Halide ions are all ligands and form a range of complexes with transition metal ions:
e.g. FeF_6^{3-}; $CuCl_4^{2-}$

Anomalous or unusual properties of the first member of the group, fluorine

1. Hydrogen fluoride is a weak acid because the hydrogen–fluorine bond is stronger than the bonds formed by the other halogens to hydrogen.
2. Silver fluoride is very soluble. Use an enthalpy cycle to explain this (see p. 51).

Precipitation

All the halides (apart from fluorine) form insoluble silver salts; these darken in colour and get more insoluble going down the group. The change in solubility can be demonstrated by adding ammonia to the precipitated halide.

$$Ag^+(aq) + Cl^-(aq) \rightarrow AgCl(s) \text{ white; dissolves in dilute aqueous ammonia}$$
$$Ag^+(aq) + Br^-(aq) \rightarrow AgBr(s) \text{ cream: dissolves in concentrated aqueous ammonia}$$
$$Ag^+(aq) + I^-(aq) \rightarrow AgI(s) \text{ yellow: insoluble in any ammonia solution}$$

Group 7 chemistry: key ideas

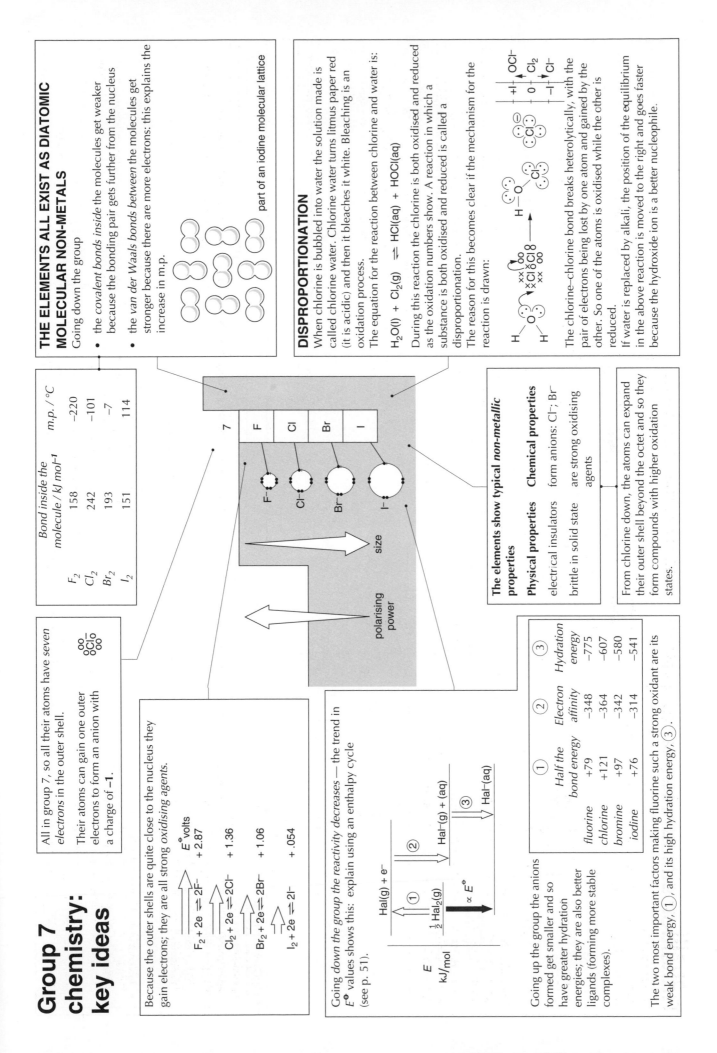

THE ELEMENTS ALL EXIST AS DIATOMIC MOLECULAR NON-METALS

Going down the group

- the *covalent bonds inside* the molecules get weaker because the bonding pair gets further from the nucleus

- the *van der Waals bonds between* the molecules get stronger because there are more electrons: this explains the increase in m.p.

part of an iodine molecular lattice

	Bond inside the molecule / kJ mol^{-1}	m.p. / °C
F_2	158	−220
Cl_2	242	−101
Br_2	193	−7
I_2	151	114

DISPROPORTIONATION

When chlorine is bubbled into water the solution made is called chlorine water. Chlorine water turns litmus paper red (it is acidic) and then it bleaches it white. Bleaching is an oxidation process.

The equation for the reaction between chlorine and water is:

$$H_2O(l) + Cl_2(g) \rightleftharpoons HCl(aq) + HOCl(aq)$$

During this reaction the chlorine is both oxidised and reduced as the oxidation numbers show. A reaction in which a substance is both oxidised and reduced is called a disproportionation.

The reason for this becomes clear if the mechanism for the reaction is drawn:

The chlorine–chlorine bond breaks heterolytically, with the pair of electrons being lost by one atom and gained by the other. So one of the atoms is oxidised while the other is reduced.

If water is replaced by alkali, the position of the equilibrium in the above reaction is moved to the right and goes faster because the hydroxide ion is a better nucleophile.

	+I	OCl⁻	
	0	Cl₂	
	−I	Cl⁻	

7

F Cl Br I

size

polarising power

The elements show typical *non-metallic* properties

Physical properties	Chemical properties
electrical insulators	form anions: Cl⁻; Br⁻
brittle in solid state	are strong oxidising agents

From chlorine down, the atoms can expand their outer shell beyond the octet and so they form compounds with higher oxidation states.

All in group 7, so all their atoms have *seven electrons* in the outer shell.

Their atoms can gain one outer electrons to form an anion with a charge of –1.

Because the outer shells are quite close to the nucleus they gain electrons; they are all strong *oxidising agents*.

E°volts

$F_2 + 2e \rightleftharpoons 2F^-$ + 2.87

$Cl_2 + 2e \rightleftharpoons 2Cl^-$ + 1.36

$Br_2 + 2e \rightleftharpoons 2Br^-$ + 1.06

$I_2 + 2e \rightleftharpoons 2I^-$ + .054

Going *down the group the reactivity decreases* — the trend in E° values shows this: explain using an enthalpy cycle (see p. 51).

$\frac{1}{2} Hal_2(g)$ ① $Hal(g) + e^-$ ② $Hal^-(g) + (aq)$ ③ $Hal^-(aq)$

E kJ/mol

$\propto E^{\ominus}$

Going up the group the anions formed get smaller and so have greater hydration energies; they are also better ligands (forming more stable complexes).

	Half the bond energy ①	Electron affinity ②	Hydration energy ③
fluorine	+79	−348	−775
chlorine	+121	−364	−607
bromine	+97	−342	−580
iodine	+76	−314	−541

The two most important factors making fluorine such a strong oxidant are its weak bond energy, ①, and its high hydration energy, ③.

Transition elements: the facts I

DEFINITIONS

d block element: any element with its highest energy electron in a d orbital

Transition element: those elements having ions with electrons in an incomplete d shell
i.e. from titanium to copper

TYPICAL PHYSICAL PROPERTIES:

all metals
high m.p.   *chromium 2160K, iron 1800K compared with sodium 371K*

hard
dense   *but not titanium*
similar atomic and ionic sizes and ionisation energies

Unlike elements in the s and p blocks, there is little change in atomic or ionic radii as the d block is crossed. This is because the additional electrons are going into an inner d sub-shell. This also results in only a small increase in ionisation energy across the d block. Although each successive nucleus has one more proton, this extra positive charge is partly shielded from the outer 4s electrons by the extra d electron in an underlying shell.

Sc^{3+}: [Ar] $3d^0 4s^0$
No d electrons
∴ not transitional

Zn^{2+}: [Ar] $3d^{10}$
full d shells
∴ not transitional

Cu^{2+}: [Ar] $3d^9$
incomplete d shells
∴ transitional

Ti	V	Cr	Mn	Fe	Co	Ni	Cu

d – block

transition elements

TYPICAL CHEMICAL PROPERTIES

1. Variable valency
Transition elements show many oxidation states; these fall into two kinds:

- **higher oxidation states: the covalently bonded oxo-compounds**

  e.g. CrO_4^{2-}; $Cr_2O_7^{2-}$; MnO_4^-; MnO_4^{2-}

- **lower oxidation states: the atomic ions**

  e.g. Cr^{3+}; Cr^{2+}; Mn^{3+}; Mn^{2+}; Fe^{3+}; Fe^{2+}; Cu^{2+}; Cu^+

2. Coloured compounds
Many of the compounds of the transition elements are coloured.

- **common examples:**

  CrO_4^{2-} – yellow; $Cr_2O_7^{2-}$ – orange; Cr^{3+} – green; Cr^{2+} – blue
  chromate (VI)   *dichromate*

  MnO_4^- – purple; MnO_4^{2-} – green; Mn^{2+} – pale pink
  manganate (VII)   *manganate (VI)*

  Fe^{3+} – yellow; Fe^{2+} – green

  Co^{2+} – pink in water, blue when dry

  Cu^{2+} – blue

3. Catalytic properties
Transition metals and their compounds can be:

- **heterogeneous catalysts, for example:**

  iron in the Haber process

  V_2O_5 in the Contact process

- **homogeneous catalysts, for example:**

  Mn^{2+} in the reaction between ethanedioate and manganate (VII)

  Fe^{2+}/Fe^{3+} in the reaction between iodide ions and peroxydisulphate (VI).

Transition elements: the facts II

Magnetic properties

Some of the transition elements are ferromagnetic which means that they they can be magnetised, e.g. iron, cobalt, and nickel.

Some of their compounds are paramagnetic which means that they move in a strong magnetic field.

The complex ion is in equilibrium with the particles which make it up and a modified equilibrium constant called the stability constant, K_{stab}, can be written for the system.

e.g. $Fe^{3+}(aq) + 6CN^-(aq) \rightleftharpoons Fe(CN)_6^{3-}(aq)$

and $K_{stab} = \dfrac{[Fe(CN)_6^{3-}(aq)]}{[Fe^{3+}(aq)] \times [CN^-(aq)]^6}$

Transition metal ions form many complex ions which vary in charge, shape, colour, and stability.

Colour

The colour of the complexes is affected by the nature of the ligand and the number of ligands around the central cation.

TYPICAL CHEMICAL PROPERTIES

4. Complex ion formation

A complex ion consists of a central ion or atom surrounded by other particles called ligands. A ligand is a particle (ion or molecule) with a lone pair which forms a dative covalent bond to the central particle. The ligands are said to be coordinated to the central particle.

dative bond forming
ligands

Charge

The charge of the complex depends on the relative charges of the central ion or atom and the ligands, and on the number of ligands around it. Complex ions may be cations or anions

cationic complexes

$Cu(NH_3)_4(H_2O)_2^{2+}(aq)$,

$FeCNS^{2+}(aq)$

anionic complexes

$CuCl_4^{2-}(aq)$,

$Fe(CN)_6^{3-}(aq)$

Shape

Ligands differ in size and this means that the number which can fit around the central cation changes. Ammonia and water are relatively small ligands and six of each can fit around a cobalt or copper ion forming octahedral complexes, while only four of the larger chloride ion can fit around either the cobalt or copper ions.

Silver is unusual in forming linear complexes.

$[NC - Ag - CN]^-$

$[H_3N - Ag - NH_3]^+$

$[O_3S_2 - Ag - S_2O_3]^{3-}$

octahedral

tetrahedral

Transition elements: the theories I

Ti	V	Cr	Mn	Fe	Co	Ni	Cu

PHYSICAL PROPERTIES

The physical properties are dominated by the fact that the electrons with the highest energies go into an inner 3d orbital rather than the outer 4s orbital. These electrons in an underlying d orbital increase the electron repulsion on the outer 4s electrons.

The elements are hard and have high m.p.s because they have high lattice energies.

The lattice energies are high because the effective nuclear charge of the cations in the lattice is high, because the electrons in the d orbitals are bad at shielding the nuclear charge.

The sizes of the atoms and ions do not decrease much as the d block is crossed. Although the atoms of each successive element have one more proton in the nucleus, increasing the attraction on the outer 4s electrons, there is increased repulsion on these outer electrons caused by the new electron in the inner d shell.

The first ionisation energies do not increase very much for the same reason. There are more protons attracting the electrons as one crosses the d block, but also more inner electrons repelling the outer electrons. The two effects almost cancel out.

CHEMICAL PROPERTIES

1. Variable valencies

In *higher oxidation states* the transition elements form molecules and molecule ions because of the *availability of vacant d and p orbitals*, which can accept electrons from the surrounding atoms.

The *lower oxidation states* happen because the transition elements have *successive ionisation energies* which are similar in value to the size of hydration energies. This is not the case with the s block metals.

e.g. compare sodium and iron

Sodium

sharp increase after 1 electron removed

electrons removed

log I.E.

Iron

steady increase in ionisation energies

electrons removed

log I.E.

So the hydration energy in solutions (and the lattice energy in solids) makes up for the slightly higher ionisation energies in the higher oxidation states of iron, but not the very much higher ionisation energies in the likes of sodium.

2. Colour

Light falling on transition element compounds interacts with the d electrons. Some of the wavelengths in the light are absorbed leaving the complementary colours to be seen.

white light

one wavelength absorbed

E ΔE E

unexcited   excited atom

3. Catalytic properties

Heterogeneous catalysis

The fact that the d block elements have 3d as well as 4s electrons helps them to form bonds with gaseous particles and so adsorb them onto the catalyst surface. This adsorption weakens the bonds in the gas particle, so lowering the activation energy of the reaction (see page 57 on catalysis).

Homogeneous catalysis

The fact that d block elements can exist in so many oxidation states is a crucial factor in making them such good homogeneous catalysts (see page 57 on catalysis).

Transition elements: the theories II

CHEMICAL PROPERTIES
4. Complex formation

d electrons are not as good at shielding the positive charge of the nucleus as either s or p electrons. This means that transition metal atoms and ions have greater polarising power than the atoms and ions from the s block.

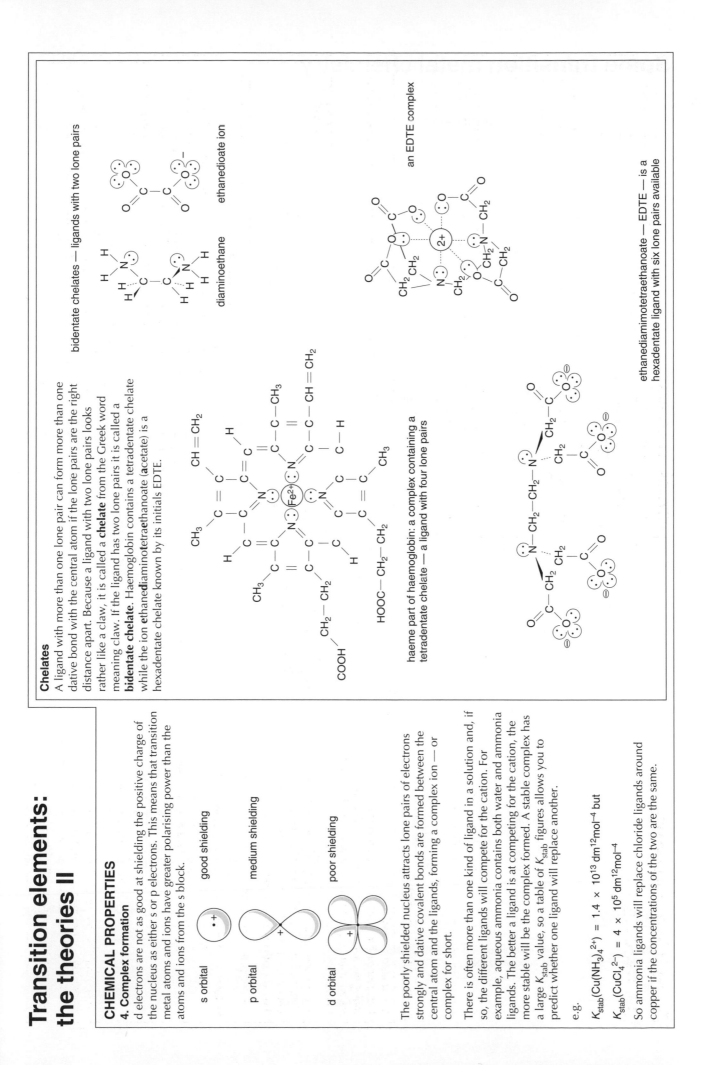

s orbital   good shielding

p orbital   medium shielding

d orbital   poor shielding

The poorly shielded nucleus attracts lone pairs of electrons strongly and dative covalent bonds are formed between the central atom and the ligands, forming a complex ion — or complex for short.

There is often more than one kind of ligand in a solution and, if so, the different ligands will compete for the cation. For example, aqueous ammonia contains both water and ammonia ligands. The better a ligand is at competing for the cation, the more stable will be the complex formed. A stable complex has a large K_{stab} value, so a table of K_{stab} figures allows you to predict whether one ligand will replace another.

e.g.

$K_{stab}(\text{Cu(NH}_3)_4{}^{2+}) = 1.4 \times 10^{13} \text{ dm}^{12}\text{mol}^{-4}$ but

$K_{stab}(\text{CuCl}_4{}^{2-}) = 4 \times 10^5 \text{ dm}^{12}\text{mol}^{-4}$

So ammonia ligands will replace chloride ligands around copper if the concentrations of the two are the same.

Chelates

A ligand with more than one lone pair can form more than one dative bond with the central atom if the lone pairs are the right distance apart. Because a ligand with two lone pairs looks rather like a claw, it is called a **chelate** from the Greek word meaning claw. If the ligand has two lone pairs it is called a **bidentate chelate**. Haemoglobin contains a tetradentate chelate while the ion **ethanediaminotetraethanoate (acetate)** is a hexadentate chelate known by its initials EDTE.

bidentate chelates — ligands with two lone pairs

ethanedioate ion

diaminoethane

an EDTE complex

haeme part of haemoglobin: a complex containing a tetradentate chelate — a ligand with four lone pairs

ethanediamimotetraethanoate — EDTE — is a hexadentate ligand with six lone pairs available

Some transition metal chemistry

VANADIUM

OX. STATE

+V	VO_2^+ yellow	
	VO_2^+ / VO^{2+} $E^{\ominus} = +1.00V$	
+IV	VO^{2+} blue	
	VO^{2+} / V^{3+} $E^{\ominus} = +0.34V$	
+III	V^{3+} green	
	V^{3+} / V^{2+} $E^{\ominus} = +0.26V$	
+II	V^{2+} purple	

REDUCTION OF VO_2^+

To VO^{2+}: redox couple with $E^{\ominus}$ between 1 and +0.34 volts i.e. I_2/I^- $E^{\ominus} = 0.54V$ so add I^- (aq)

To V^{3+}: redox couple with $E^{\ominus}$ between +0.34 and −0.26V e.g. Sn^{2+}/Sn $E^{\ominus} = -0.14V$ so add Sn metal

To V^{2+}: redox couple with $E^{\ominus}$ less than −0.26V e.g. Zn^{2+}/Zn $E^{\ominus} = -0.76V$ so add Zn metal

		Sc	Ti	V	Cr	Mn

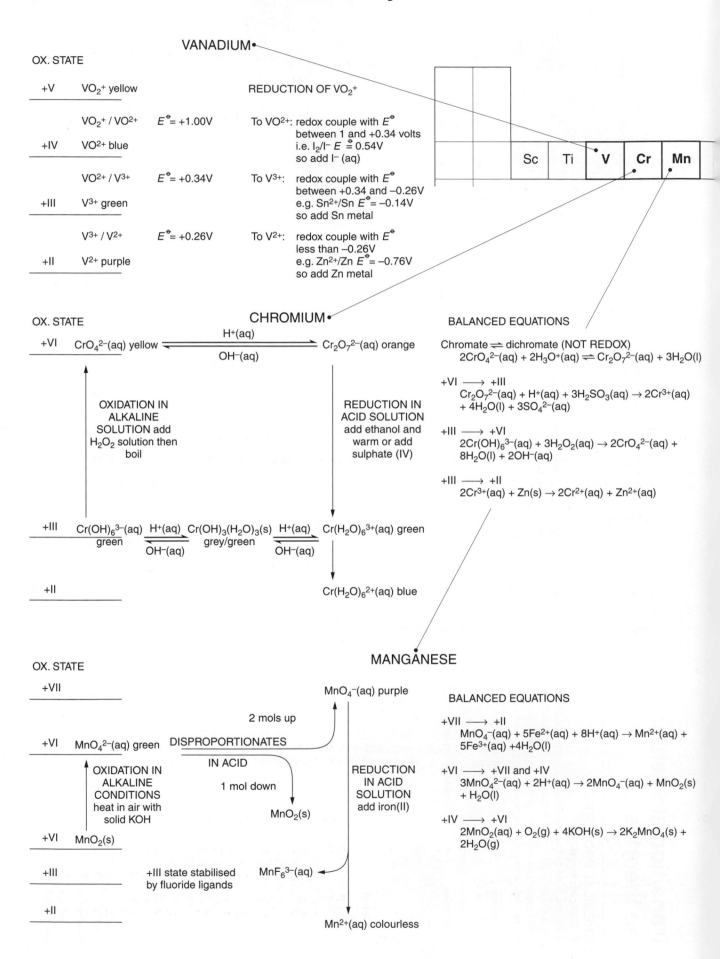

CHROMIUM

OX. STATE

| +VI | CrO_4^{2-}(aq) yellow | $\xrightleftharpoons[OH^-(aq)]{H^+(aq)}$ | $Cr_2O_7^{2-}$(aq) orange |

OXIDATION IN ALKALINE SOLUTION add H_2O_2 solution then boil

REDUCTION IN ACID SOLUTION add ethanol and warm or add sulphate (IV)

| +III | $Cr(OH)_6^{3-}$(aq) green | $\xrightleftharpoons[OH^-(aq)]{H^+(aq)}$ | $Cr(OH)_3(H_2O)_3$(s) grey/green | $\xrightleftharpoons[OH^-(aq)]{H^+(aq)}$ | $Cr(H_2O)_6^{3+}$(aq) green |

| +II | | | $Cr(H_2O)_6^{2+}$(aq) blue |

BALANCED EQUATIONS

Chromate $\rightleftharpoons$ dichromate (NOT REDOX)
$$2CrO_4^{2-}(aq) + 2H_3O^+(aq) \rightleftharpoons Cr_2O_7^{2-}(aq) + 3H_2O(l)$$

+VI $\longrightarrow$ +III
$$Cr_2O_7^{2-}(aq) + H^+(aq) + 3H_2SO_3(aq) \rightarrow 2Cr^{3+}(aq) + 4H_2O(l) + 3SO_4^{2-}(aq)$$

+III $\longrightarrow$ +VI
$$2Cr(OH)_6^{3-}(aq) + 3H_2O_2(aq) \rightarrow 2CrO_4^{2-}(aq) + 8H_2O(l) + 2OH^-(aq)$$

+III $\longrightarrow$ +II
$$2Cr^{3+}(aq) + Zn(s) \rightarrow 2Cr^{2+}(aq) + Zn^{2+}(aq)$$

MANGANESE

OX. STATE

| +VII | | MnO_4^-(aq) purple |
| +VI | MnO_4^{2-}(aq) green | DISPROPORTIONATES |

2 mols up

IN ACID

1 mol down

MnO_2(s)

OXIDATION IN ALKALINE CONDITIONS heat in air with solid KOH

REDUCTION IN ACID SOLUTION add iron(II)

+VI	MnO_2(s)	
+III		+III state stabilised by fluoride ligands	MnF_6^{3-}(aq)
+II		Mn^{2+}(aq) colourless

BALANCED EQUATIONS

+VII $\longrightarrow$ +II
$$MnO_4^-(aq) + 5Fe^{2+}(aq) + 8H^+(aq) \rightarrow Mn^{2+}(aq) + 5Fe^{3+}(aq) + 4H_2O(l)$$

+VI $\longrightarrow$ +VII and +IV
$$3MnO_4^{2-}(aq) + 2H^+(aq) \rightarrow 2MnO_4^-(aq) + MnO_2(s) + H_2O(l)$$

+IV $\longrightarrow$ +VI
$$2MnO_2(aq) + O_2(g) + 4KOH(s) \rightarrow 2K_2MnO_4(s) + 2H_2O(g)$$

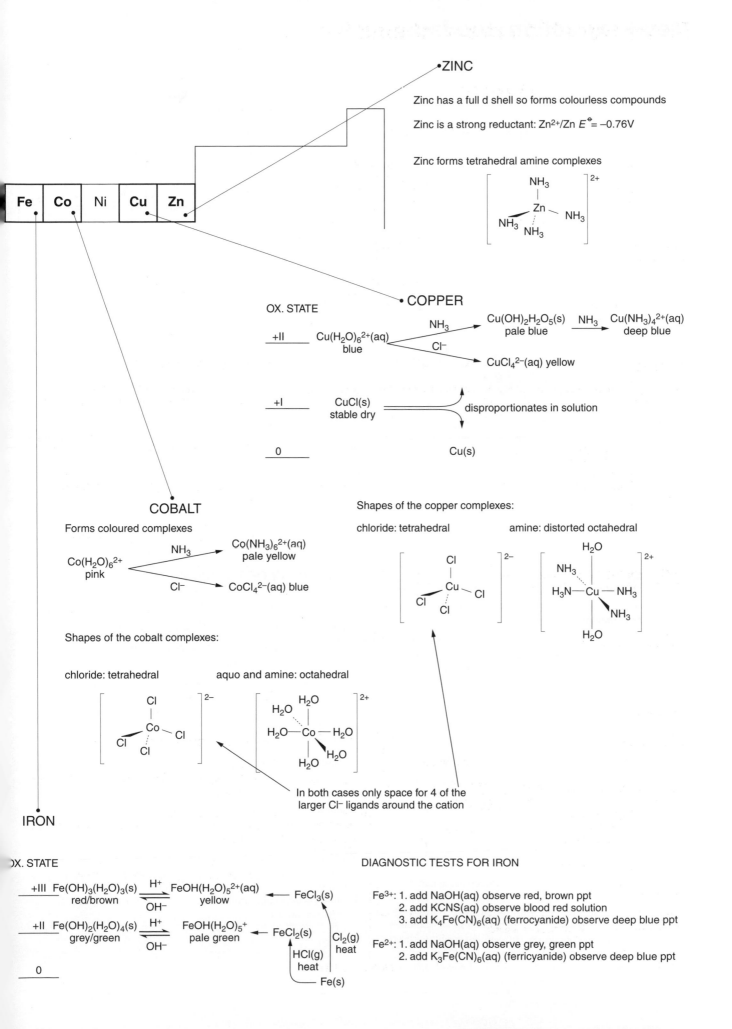

ZINC

Zinc has a full d shell so forms colourless compounds

Zinc is a strong reductant: Zn^{2+}/Zn $E^{\ominus} = -0.76V$

Zinc forms tetrahedral amine complexes

$$\left[\begin{array}{c} NH_3 \\ | \\ H_3N \cdots Zn - NH_3 \\ | \\ NH_3 \end{array} \right]^{2+}$$

COPPER

OX. STATE

+II $\quad$ $Cu(H_2O)_6^{2+}(aq)$ $\xrightarrow{NH_3}$ $Cu(OH)_2H_2O_5(s)$ $\xrightarrow{NH_3}$ $Cu(NH_3)_4^{2+}(aq)$
$\qquad$ blue $\qquad\qquad\qquad$ pale blue $\qquad\qquad\qquad$ deep blue
$\qquad\qquad\qquad\qquad \xrightarrow{Cl^-} CuCl_4^{2-}(aq)$ yellow

+I $\quad$ $CuCl(s)$ stable dry $\qquad$ disproportionates in solution

0 $\qquad\qquad\qquad\qquad\qquad$ $Cu(s)$

COBALT

Forms coloured complexes

$Co(H_2O)_6^{2+}$ $\xrightarrow{NH_3}$ $Co(NH_3)_6^{2+}(aq)$ pale yellow
pink $\qquad \xrightarrow{Cl^-} CoCl_4^{2-}(aq)$ blue

Shapes of the cobalt complexes:

chloride: tetrahedral $\qquad\qquad$ aquo and amine: octahedral

$$\left[\begin{array}{c} Cl \\ | \\ Cl \cdots Co - Cl \\ | \\ Cl \end{array} \right]^{2-} \qquad \left[\begin{array}{c} H_2O \quad H_2O \\ H_2O - Co - H_2O \\ H_2O \quad H_2O \end{array} \right]^{2+}$$

In both cases only space for 4 of the larger Cl^- ligands around the cation

Shapes of the copper complexes:

chloride: tetrahedral $\qquad\qquad$ amine: distorted octahedral

$$\left[\begin{array}{c} Cl \\ | \\ Cl \cdots Cu - Cl \\ | \\ Cl \end{array} \right]^{2-} \qquad \left[\begin{array}{c} H_2O \\ NH_3 \\ H_3N - Cu - NH_3 \\ NH_3 \\ H_2O \end{array} \right]^{2+}$$

IRON

OX. STATE

+III $\quad$ $Fe(OH)_3(H_2O)_3(s)$ $\underset{OH^-}{\overset{H^+}{\rightleftharpoons}}$ $FeOH(H_2O)_5^{2+}(aq)$ $\leftarrow$ $FeCl_3(s)$
$\qquad$ red/brown $\qquad\qquad\qquad\qquad$ yellow

+II $\quad$ $Fe(OH)_2(H_2O)_4(s)$ $\underset{OH^-}{\overset{H^+}{\rightleftharpoons}}$ $FeOH(H_2O)_5^+$ $\leftarrow$ $FeCl_2(s)$ $\begin{array}{c} Cl_2(g) \\ heat \end{array}$
$\qquad$ grey/green $\qquad\qquad\qquad\qquad$ pale green $\qquad\qquad$ $\begin{array}{c} HCl(g) \\ heat \end{array}$

0 $\qquad\qquad\qquad\qquad\qquad\qquad\qquad\qquad\qquad$ $Fe(s)$

DIAGNOSTIC TESTS FOR IRON

Fe^{3+}: 1. add NaOH(aq) observe red, brown ppt
$\qquad$ 2. add KCNS(aq) observe blood red solution
$\qquad$ 3. add $K_4Fe(CN)_6$(aq) (ferrocyanide) observe deep blue ppt

Fe^{2+}: 1. add NaOH(aq) observe grey, green ppt
$\qquad$ 2. add $K_3Fe(CN)_6$(aq) (ferricyanide) observe deep blue ppt

Metal extraction: the facts

EXTRACTION OF ALUMINIUM

Input

bauxite, $Al_2O_3.3H_2O$, the source of aluminium

cryolite, Na_3AlF_6, added to lower the m.p. of the purified aluminium oxide

sodium hydroxide to purify the bauxite

carbon, from crude oil, to make anodes

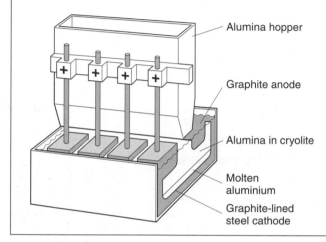

- Alumina hopper
- Graphite anode
- Alumina in cryolite
- Molten aluminium
- Graphite-lined steel cathode

The process

The bauxite is purified by dissolving in sodium hydroxide. The aluminium oxide dissolves because it is amphoteric, and can be filtered off leaving the impurities — iron oxide, silica, and titanium oxide. The aluminate solution is seeded with pure aluminium oxide and precipitates out pure oxide. This is filtered off and the sodium hydroxide recycled.

The pure oxide is dried and mixed with cryolite. This lowers the m.p. from 2045 to 950 °C.

The oxide is melted and electrolysed in a carbon lined tank acting as the cathode. Aluminium is produced at the cathode and sinks to the bottom from where it is siphoned off at intervals. The oxide ions are oxidised at the carbon anodes suspended in the melt. Oxygen gas is produced which reacts with the hot anodes making carbon dioxide. There are several anodes in each cell and these are replaced in turn as they get burnt away.

Cathode reaction

$$Al^{3+}(l) + 3e^- \rightarrow Al(l)$$

Anode reactions

$$2O^{2-}(l) \rightarrow O_2(g) + 4e^- \text{ then } O_2(g) + C(s) \rightarrow CO_2(g)$$

EXTRACTION OF CHROMIUM

Chromium oxide, Cr_2O_3, is reduced chemically in a batch process, usually using a more reactive metal, aluminium, in a thermite type reaction.

The oxide is mixed with aluminium powder and ignited in a refractory lined container using a fuse of magnesium and barium peroxide. A very exothermic reaction takes place producing 97–99% pure metal.

$$Cr_2O_3(s) + 2Al(s) \rightarrow 2Cr(s) + Al_2O_3(s) \qquad \Delta H = -536 \text{ kJ}$$

Chromium and nickel are alloyed with iron to make stainless steel. Although chromium is more reactive than iron it is, like aluminium, protected by an oxide layer which stops corrosion.

EXTRACTION OF ZINC

Like iron, zinc is extracted in a blast furnace using carbon as the reducing agent. Unlike iron, zinc has a low boiling point and it comes off the top of the blast furnace as a vapour. This vapour is cooled by passing it through a fine spray of molten lead in a lead splash condenser. This sudden cooling prevents the hot zinc vapour reoxidising. The mixture of lead and zinc is then cooled, the solubility of the zinc in the lead falls, and the zinc crystallises out and is extracted.

EXTRACTION OF TITANIUM

Titanium (IV) oxide is converted to the liquid chloride by heating it with carbon and chlorine.

$$TiO_2(s) + 2Cl_2(g) + 2C(s) \rightarrow TiCl_4(g) + 2CO(g)$$

This is then reduced chemically using a very reactive metal. Either magnesium or sodium is used.

In the magnesium process, bars of magnesium are heated to 700 °C in a steel vessel under an argon atmosphere. Liquid titanium (IV) chloride is then poured in and a very exothermic reaction takes place. The temperature is kept between 850 and 900 °C

$$TiCl_4(g) + 2Mg(l) \rightarrow Ti(s) + 2MgCl_2(l) \qquad \Delta H = -540 \text{ kJ}$$

The magnesium chloride is tapped off, and electrolysed, recycling both the magnesium and the chlorine. Impure titanium is made which can be refined by remelting it under a noble gas in an electric arc furnace.

Titanium has the same mechanical strength as steel, but is much less dense. It has a very high m.p. of 1675°C and is chemically very unreactive at room temperature. Its main uses are in space and aeronautical engineering where its lightness is important, but the high cost of extraction prevents its widespread use.

Iron and steel

EXTRACTION OF IRON: THE BLAST FURNACE

The charge is loaded into the top of the blast furnace. This contains:

1. iron ore: the source of iron
2. coke: the fuel and reducing agent
3. limestone: to form a slag to dissolve high m.p. impurities.

Preheated air is blasted into the bottom of the furnace.

The key reactions are:

1. The coke burns

$$C(s) + O_2(g) \rightarrow CO_2(g)$$

2. More coke reduces carbon dioxide making carbon monoxide

$$C(s) + CO_2(g) \rightleftharpoons 2CO_2(g)$$

3. Carbon monoxide moving up the furnace reduces iron ore falling down the furnace

$$Fe_2O_3(s) + 3CO(g) \rightarrow 2Fe(l) + 3CO_2(g)$$

4. Unreacted iron ore is reduced by unreacted coke

$$2Fe_2O_3(s) + 3C(s) \rightarrow 4Fe(l) + 3CO_2(g)$$

5. Limestone decomposes

$$CaCO_3(s) \rightarrow CaO(s) + CO_2(g)$$

6. Calcium oxide reacts with high m.p. impurities forming a slag, which also protects the molten iron from the air blast

$$CaO(s) + SiO_2(s) \rightarrow CaSiO_3(l)$$

The crude iron (pig iron) from the blast furnace contains the following impurities:
carbon, from the coke
nitrogen, from the air blast
silicon, from sandy impurities in the ore

These make the iron very brittle and have to be removed. This is done in a converter as the iron is turned into steel

STEEL MAKING

Preheated oxygen is blown through the molten iron, burning out all the impurities. Because this reaction is very exothermic, scrap steel is also added to keep the temperature down

Weighed amounts of carbon and other alloying elements are then added:
chromium and nickel for stainless steel
manganese for spring steel
tungsten for cutting steel

RUSTING

Aluminium and iron are the two most important industrial metals. Aluminium is very strong for its weight — it has a high strength/weight ratio — and is protected by its oxide layer, but it is more expensive to extract. Iron is cheaper, but its greatest disadvantage is that in the presence of air and water it rusts. Rusting is an electrochemical process. The metal lattice varies due to impurities and stress so some of the atoms have higher energies than the others. The water in contact with the iron also varies in the amount of dissolved oxygen and other impurities it contains. These differences result in cells being set up as different parts of the metal have different electrode potentials. Some areas are anodic as iron atoms are oxidised into ions and go into solution:

$$Fe(s) \rightarrow Fe^{2+}(aq) + 2e^- \quad E^{\ominus} = -0.44 \text{ volts}$$

The electrons released flow through the metal to a cathodic area where dissolved oxygen is reduced.

$$4e^- + O_2(g) + 2H_2O(l) \rightarrow 4OH^-(aq)$$
$$E^{\ominus} = +0.40 \text{ volts}$$

The ions from these two reactions diffuse together and form a precipitate:

$$Fe^{2+}(aq) + 2OH^-(aq) \rightarrow Fe(OH)_2(s)$$

This precipitate is oxidised by dissolved oxygen forming rust.

$$4Fe(OH)_2(s) + O_2(g) + 2H_2O(l) \rightarrow 4Fe(OH)_3(s)$$

Preventing rust

- Keep the water and air away from the iron: grease, paint, coat, plate, or enamel the metal.

- Make the metal less reactive: alloy with a less reactive metal like chromium.

- Sacrificial protection: connect a more reactive metal — Zn — which reacts in preference.

Metal extraction: the theory

KEY FACTORS

The way in which a metal is extracted depends on:

1. the abundance of the metal
2. the reactivity of the metal
3. the demand for the metal and the purity that is required

THE ABUNDANCE OF METALS

Some metals such as sodium, calcium, iron, and aluminium are found widespread, while others like copper, chromium, and zinc are found in rich but localised deposits. Estimates have been made of the composition of the earth's crust, which show that iron and aluminium are by far the most plentiful of the industrial metals:

Metal	% in earth's crust
aluminium	8.1
iron	5.0
calcium	3.6

ELECTROLYTIC REDUCTION: THE PRINCIPLES

1. The ores are all high m.p. ionic solids. Impurities are added to lower the m.p., e.g. cryolite lowers the m.p. of bauxite from 2045 to 950 °C.

2. The amount of metal produced at the cathode depends on the current flowing so most cells operate at enormous currents.

3. The products are very reactive metals and non-metals and so the cell has to be designed to keep them apart.

4. The process is expensive because of the amount of electricity needed to keep the cell hot and reduce the ore.

REACTIVITY OF METALS

The periodic table gives an indication of the reactivity of the metals in the main groups, but it does not help with the transition metals. A good guide to the reactivity of metals is their standard electrode potentials. If some of the metals are listed in increasing $E^{\ominus}$ order we get:

potassium calcium sodium magnesium	extracted by electrolysis of molten or fused ore
aluminium titanium chromium zinc iron	extracted by chemical reduction in a furnace
copper silver	extracted by thermal decomposition

THE DEMAND FOR METALS

The two metals in greatest demand are iron and aluminium which are the two main industrial metals.

Iron is cheaper to extract. It can be converted into a range of steel alloys with very different properties.

Aluminium is lighter but stronger (it has a high strength/weight ratio) and it is protected by its oxide layer, but it is more expensive to extract. It also conducts electricity well.

Copper has many uses because of its physical properties. It can be used for pipes and sheet because it is so ductile; wires and cables because it is such a good conductor when pure; and coinage because it is relatively unreactive and makes coins a distinctive colour.

Titanium has the highest strength/weight ratio of all metals.

Zinc, being a reactive metal, is widely used in the sacrificial protection of steel structures. Steel objects are either dipped (galvanised) or have zinc anodes attached to them.

CHEMICAL REDUCTION: THE PRINCIPLES

1. The cost of the reductant is very important. Carbon (coke) is the cheapest. If the reductant is another very reactive metal, the cost of extracting it has to be considered.

2. A continuous process is always cheaper to run than a batch process, but the demand for the metal may not justify the capital cost of a continuous plant.

3. The metal may be contaminated by the reductant and need further purification, e.g. iron is turned into steel.

Organic chemistry: introduction

A good organic answer:

makes a statement
gives an example or illustrates the statement
explains it using a theory or mechanism

It will be a factual statement about *physical* or *chemical properties* which is explained by *theories* and *mechanisms* that you have learnt in physical chemistry.

e.g.	**Facts** listed with an example	**Theories** theories and ideas used
	Physical properties such as: state and fixed points physical properties solubility	explained in terms of **structure and bonding;** e.g. **chain length**
	Chemical properties such as: reactions with water, acid, or base	explained in terms of **acid–base equilibria**
	reactions with oxidising or reducing agents	explained in terms of **redox equilibria**
	reactions with nucleophiles and electrophiles	explained in terms of **functional group and its reactive site and mechanism**

SOME IMPORTANT THEMES

Physical properties in similar compounds are controlled by **chain length**.

The length of the chain controls the number of atoms and so the number of electrons and so the strength of the van der Waals forces between molecules.

Chemical properties are controlled by the kind of **links** (**functional groups**) in or joined to the chain: each functional group has particular reactive sites.

Each **reactive site** gives the substance particular properties.

Two reactive sites next to each other modify each other.

The type of carbon skeleton can change the properties of the reactive site so compare:

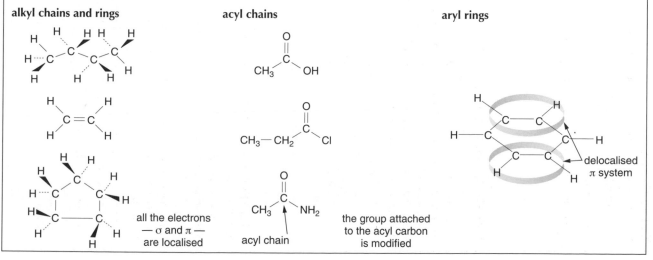

alkyl chains and rings

all the electrons
— σ and π —
are localised

acyl chains

acyl chain

the group attached
to the acyl carbon
is modified

aryl rings

delocalised
π system

Functional groups and naming in organic compounds

The hydrocarbon part of an organic molecule is fairly unreactive, so the chemistry of organic compounds is often dominated by other atoms or groups of atoms joined to the chain. These atoms or groups of atoms, whose reactions dominate the chemistry of the molecule, are called **functional groups**.

NAMING ORGANIC COMPOUNDS

The name consists of three parts:

- the *first* part tells you the *chain length*

1 carbon = **meth-**	5 carbons = **pent-**
2 carbons = **eth-**	6 carbons = **hex-**
3 carbons = **prop-**	7 carbons = **hept-**
4 carbons = **but-**	8 carbons = **oct-**

- the *second* part tells you about the linking or bonding in the chain

  -**an-** means all **single** bonds in the carbon chain

  -**en-** means a **double** bond in the carbon chain

  -**yn-** means a **triple** bond in the chain

- the *last* part tells you what *functional group is joined* to the chain

  -**e** means only **hydrogen** is joined to the chain

  -**ol** means an **-OH**

  -**amine** means an **-NH₂**

  -**al** means a **carbonyl group on the end of the chain**

  -**one** means a **carbonyl group on the chain, but not at the end**

  -**oic acid** means a **carboxylic group** on the chain

  Numbers are used to give the position along the chain.

Some common functional groups

Name of the group	Type of compound	Formula or structure
Hydroxyl	Alcohol	— OH
Halogeno	Halide	— Hal
Amino	Amine	— NH₂
Carbonyl	Aldehyde (at the end of the chain)	
	or ketone (in the middle of the chain)	
Carboxyl	Carboxylic acid	

Examples

eth-an-al
two carbons
single bond in the chain
carbonyl at the end of the chain

but-an-oic acid
four carbons
single bonds in the chain
carboxylic acid at the end

prop-2-en-1-ol
three carbons
double bond in the chain
hydroxyl group at the end

pent-an-2-one
five carbons
single bonds in the chain
carbonyl on the second carbon

(the dashes between the separate parts of the name are left out unless numbers are needed)

Bonding and structure in organic compounds

Carbon forms a huge number of compounds: about 20 times more than all the other elements put together. The study of this great number of compounds all based on carbon is called organic chemistry.

There are a number of special reasons for this behaviour of carbon:

2. QUADRA- OR TETRA-VALENCY

Carbon is in group 4 and forms four bonds.

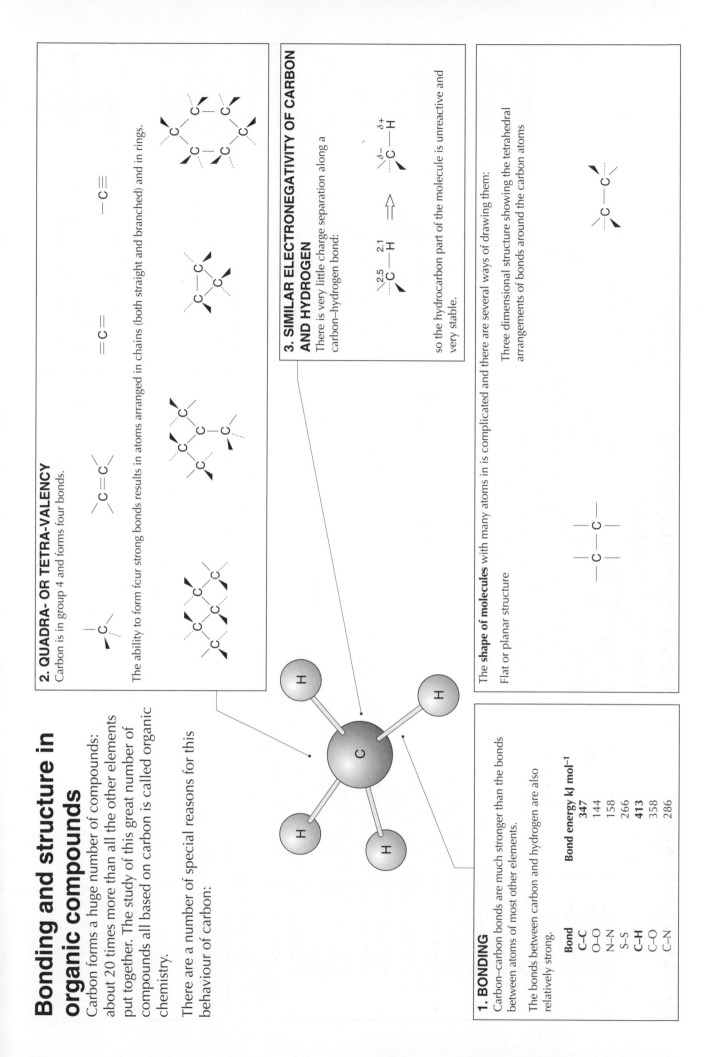

The ability to form four strong bonds results in atoms arranged in chains (both straight and branched) and in rings.

3. SIMILAR ELECTRONEGATIVITY OF CARBON AND HYDROGEN

There is very little charge separation along a carbon–hydrogen bond:

$$\overset{2.5}{C}—\overset{2.1}{H} \implies \overset{\delta-}{C}—\overset{\delta+}{H}$$

so the hydrocarbon part of the molecule is unreactive and very stable.

The **shape of molecules** with many atoms in is complicated and there are several ways of drawing them:

Three dimensional structure showing the tetrahedral arrangements of bonds around the carbon atoms

Flat or planar structure

1. BONDING

Carbon–carbon bonds are much stronger than the bonds between atoms of most other elements.

The bonds between carbon and hydrogen are also relatively strong.

Bond	Bond energy kJ mol^{-1}
C–C	**347**
O–O	144
N–N	158
S–S	266
C–H	**413**
C–O	358
C–N	286

Key ideas in organic chemistry

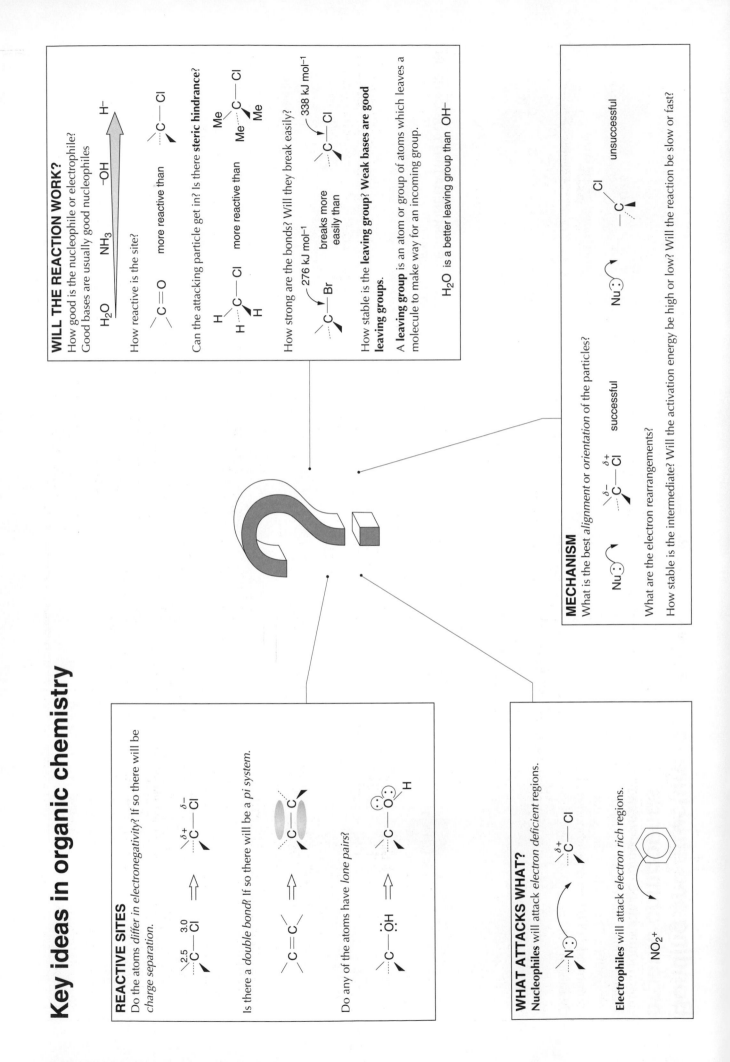

REACTIVE SITES
Do the atoms *differ in electronegativity?* If so there will be *charge separation.*

Is there a *double bond?* If so there will be a *pi system.*

Do any of the atoms have *lone pairs?*

WHAT ATTACKS WHAT?
Nucleophiles will attack *electron deficient* regions.

Electrophiles will attack *electron rich* regions.

NO_2^+

WILL THE REACTION WORK?
How good is the nucleophile or electrophile?
Good bases are usually good nucleophiles

H_2O NH_3 $-OH$ H^-

How reactive is the site?

$C=O$ more reactive than $C-Cl$

Can the attacking particle get in? Is there **steric hindrance?**

more reactive than

How strong are the bonds? Will they break easily?

276 kJ mol^{-1} C—Br   breaks more easily than 338 kJ mol^{-1} C—Cl

How stable is the **leaving group?** **Weak bases are good leaving groups.**

A **leaving group** is an atom or group of atoms which leaves a molecule to make way for an incoming group.

H_2O is a better leaving group than OH^-

MECHANISM
What is the best *alignment* or *orientation* of the particles?

successful

unsuccessful

What are the electron rearrangements?

How stable is the *intermediate?* Will the activation energy be high or low? Will the reaction be slow or fast?

Reactive sites and organic reagents

FUNCTIONAL GROUPS AND REACTIVE SITES

Functional groups control the chemistry of molecules because they contain **reactive sites**.

There are two kinds of *reactive site* in a molecule.

Electron deficient sites (shown by the δ+ sign) are short of electrons.

Electron excessive sites (shown by the δ− sign) are rich in electrons.

Electron deficient sites	**Electron excessive sites**
an atom bonded to a more electronegative one	an atom with lone pair(s)

a pi (π) system

positive ions

CLASSIFYING ORGANIC REACTIONS

Organic reactions are classified in two ways:

1. *By the type of reagent being used.* If a nucleophile is being added the reaction is called **nucleophilic**; if an electrophile is added it is called **electrophilic**.

2. *By what happens when the reaction is over.* If an atom or group has been *added* the reaction is called an **addition**, or the opposite might happen and *atoms from two neighbouring carbons are lost*, when the reaction is called an **elimination**. If one atom or group is *replaced by another*, the reaction will be called a **substitution**. In a substitution, the atom or group being replaced is called the **leaving group**. Particles which leave easily are called *good leaving groups*. Usually *weak bases* (bad nucleophiles) are *good leaving groups*.

On the next pages you will see examples of:

Nucleophilic substitution	**Electrophilic substitution**
Nucleophilic addition	**Electrophilic addition**

REAGENTS

Reagents are substances that are added to organic compounds to make them react. Typical reagents are bromine water and sodium hydroxide.

Reagents are classified by the kind of reactive site they react with.

Reagents that are attracted to regions of positive charge or electron deficient sites are called **nucleophiles**. Nucleophiles are particles with **lone pairs of electrons.**

A useful general rule is that good bases are usually good nucleophiles

Reagents that are attracted to regions of negative charge or electron excessive sites are called **electrophiles**. Sometimes a dipole is *induced* or *caused* by the compound it is reacting with, making an electrophile. So an electrophile is **an electron pair acceptor.**

$$NO_2^+ \qquad \overset{\delta-}{Br} - \overset{\delta+}{Br} \qquad \text{(induced by a neighbouring particle)}$$

REACTION MECHANISMS

During any reaction, bonds are broken and made. As bonds are attractive forces between positive and negative bits of particles, making and breaking them means moving electrons around.

A mechanism is a description of a successful collision between the reactants (a reagent particle and the organic compound particle) and the electron rearrangements that happen as reactants are changed into products.

Most mechanisms are a series of three labelled diagrams showing the particles:

- just **before** a successful collision and therefore lined up properly

- **during** the collision as the electrons are moving around.

- and **after** the collision.

Drawing mechanisms

ELECTRON REARRANGEMENTS AND CURLY ARROWS

During a reaction bonds are broken and made.

Bonds consist of electrons and so during a reaction, electrons are moving.

Electrons in the outer shell have to be:

sigma pairs between a pair of atoms   **pi pairs** between a pair of atoms   **lone pairs** on a single atom

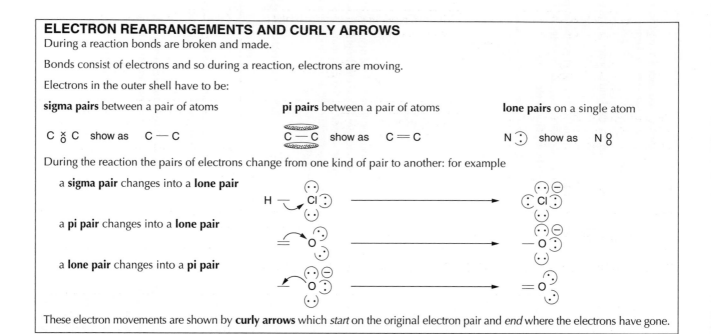

During the reaction the pairs of electrons change from one kind of pair to another: for example

a **sigma pair** changes into a **lone pair**

a **pi pair** changes into a **lone pair**

a **lone pair** changes into a **pi pair**

These electron movements are shown by **curly arrows** which *start* on the original electron pair and *end* where the electrons have gone.

A simple acid–base reaction
$H_2O + HCl \rightarrow H_3O^+ + Cl^-$

For the reaction to happen, a new bond must be made between the oxygen of the water molecule and the hydrogen atom of the hydrogen chloride. Also the bond between the hydrogen and the chlorine atom must break. These changes can be shown using a mechanism, in which the movement of the electrons is shown by curly arrows:

Here a lone pair has become a sigma pair and a sigma pair has become a lone pair. The bond between the hydrogen and chlorine has broken unevenly, with *both* the electrons ending up on the chlorine. This is an example of **heterolysis**.

Fish hooks and free radicals

A free radical forms when a covalent bond breaks in half evenly, one electron going to each atom. This kind of bond cleavage or breakage is called **homolysis**. The movement of the single electrons is shown by an arrow with half a head. This looks like, and is called, a **fish-hook**.

The particles formed have a **single unpaired electron** and are called **free radicals**.

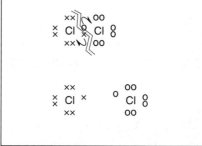

DRAWING MECHANISMS

When you are drawing the three stages of a mechanism, you must think of and comment on different things.

Stage 1: Here the most important thing to think about is getting the *two particles lined up* properly so that the oppositely charged bits are next to each other. In your labelling, you should try to explain why each bit has the charge you have shown.

Stage 2: In this part, you will be thinking about how *strong the bonds* are that might break and where the electrons are going if the bond does break. Is the particle being attacked *saturated* or not? If it is *saturated*, there is not much room for the incoming electrons and the reaction will have a *high activation energy*. This means it will go *slowly*. If the particle being attacked is *unsaturated*, then π electrons can be moved out of the way easily, becoming lone pairs or σ pairs and so the reaction will have a *low activation energy* and go quite *quickly*.

Stage 3: Here you think about the energy or stability of what you have made. Is the leaving group stable?

Heterolytic nucleophilic substitution: S_N1 and S_N2: halogenoalkanes with nucleophiles

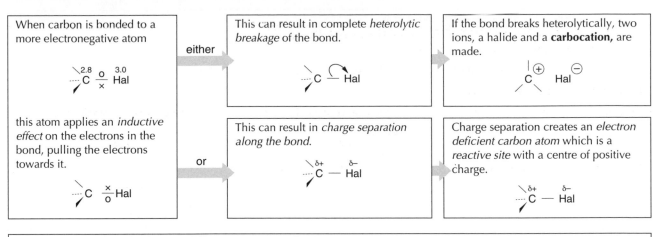

When carbon is bonded to a more electronegative atom

this atom applies an *inductive effect* on the electrons in the bond, pulling the electrons towards it.

either

This can result in complete *heterolytic breakage* of the bond.

or

This can result in *charge separation along the bond.*

If the bond breaks heterolytically, two ions, a halide and a **carbocation,** are made.

Charge separation creates an *electron deficient carbon atom* which is a *reactive site* with a centre of positive charge.

These two possibilities lead to two different reaction mechanisms:

SUBSTITUTION BY A NUCLEOPHILE; FIRST ORDER KINETICS: S_N1

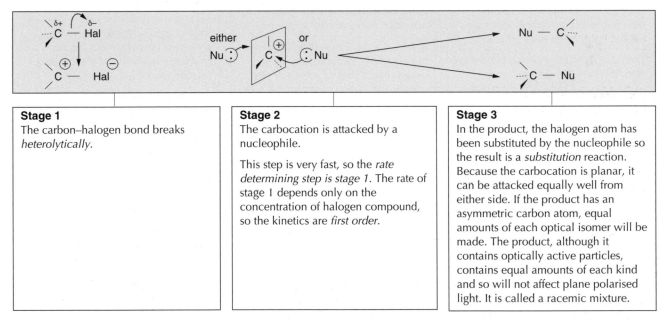

Stage 1
The carbon–halogen bond breaks *heterolytically.*

Stage 2
The carbocation is attacked by a nucleophile.

This step is very fast, so the *rate determining step is stage 1*. The rate of stage 1 depends only on the concentration of halogen compound, so the kinetics are *first order.*

Stage 3
In the product, the halogen atom has been substituted by the nucleophile so the result is a *substitution* reaction. Because the carbocation is planar, it can be attacked equally well from either side. If the product has an asymmetric carbon atom, equal amounts of each optical isomer will be made. The product, although it contains optically active particles, contains equal amounts of each kind and so will not affect plane polarised light. It is called a racemic mixture.

SUBSTITUTION BY A NUCLEOPHILE; SECOND ORDER KINETICS: S_N2

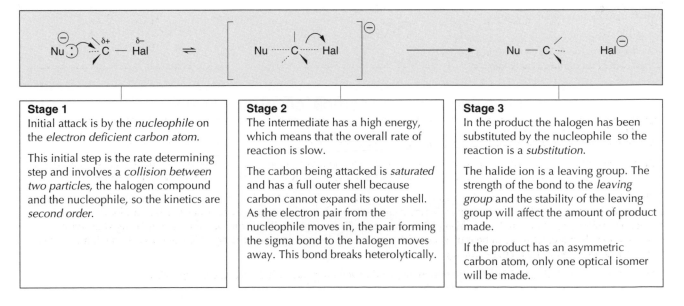

Stage 1
Initial attack is by the *nucleophile* on the *electron deficient carbon atom.*

This initial step is the rate determining step and involves a *collision between two particles*, the halogen compound and the nucleophile, so the kinetics are *second order.*

Stage 2
The intermediate has a high energy, which means that the overall rate of reaction is slow.

The carbon being attacked is *saturated* and has a full outer shell because carbon cannot expand its outer shell. As the electron pair from the nucleophile moves in, the pair forming the sigma bond to the halogen moves away. This bond breaks heterolytically.

Stage 3
In the product the halogen has been substituted by the nucleophile so the reaction is a *substitution.*

The halide ion is a leaving group. The strength of the bond to the *leaving group* and the stability of the leaving group will affect the amount of product made.

If the product has an asymmetric carbon atom, only one optical isomer will be made.

Heterolytic nucleophilic addition and addition/elimination: carbonyl compounds with hydrogen cyanide

Hydrogen cyanide adds across the double bond in a carbonyl compound, if there is a trace of the catalyst potassium cyanide present.

$$\text{\Large$\diagdown$}C=O \;\; + \;\; H-CN \;\longrightarrow\; \underset{NC}{\text{\Large$\diagdown$}}C-O{\overset{H}{}}$$

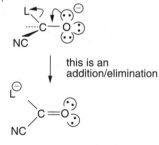

Stage 1

Initial attack is by the lone pair of a nucleophile on an electron deficient carbon atom. The hydrogen cyanide is not a good nucleophile, because it does not have a prominent lone pair. But the cyanide ion does.

The positive reactive site on the carbon atom is due to the electron withdrawing or *inductive effect* of the more electronegative oxygen atom bonded to it.

The carbon atom is *unsaturated*. As the lone pair from the nucleophile moves in, the π pair between the carbon and the oxygen moves onto the oxygen at the same time. This movement of π electrons is called a **mesomeric shift** and is the heterolytic breaking of the π bond.

Stage 2

The intermediate made is a four bonded carbon atom. This is the usual bonding for carbon and so the energy of the intermediate is low. A low activation energy means a fast reaction.

If there is a good leaving group joined to the carbon, the lone pair on the oxygen may move back between the oxygen and carbon making the π bond again as the leaving group moves away.

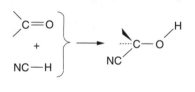

this is an addition/elimination

Stage 3

The negatively charged oxygen atom made when the π pair moves onto it (becoming a lone pair) is a good base.

This base deprotonates the protonated form of the nucleophile, here the hydrogen cyanide, replacing the cyanide ion which did the first attack. So the cyanide ion is a catalyst.

Notice that the result of the whole reaction is the *addition* of a substance across the double bond. The hydrogen always ends up joined to the oxygen.

Because the carbonyl group is planar, the nucleophile has an equal chance of attacking from above or below the plane of the molecule.

This means that there are two possible products made in equal quantities.

The two products can be optical isomers. As they are made in equal quantities, the product mixture is a racemate.

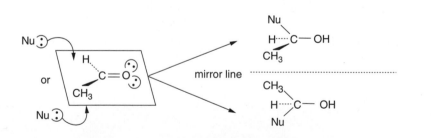

HETEROLYTIC NUCLEOPHILIC ADDITION: GRIGNARD REAGENTS

Use of Grignard reagents yields the unusual nucleophilic carbon ion. Typically a Grignard reagent is made by refluxing a halogenoalkane with magnesium turnings in dry ether. The resulting covalent metal compound yields ions:

i.e. $\quad R-Br + Mg \rightleftharpoons R-Mg-Br \rightleftharpoons R^- \overset{+}{M}gBr$

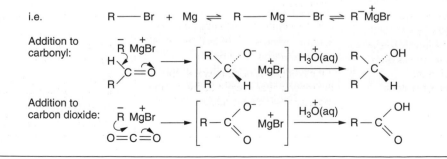

Heterolytic electrophilic substitution:
the nitration of benzene

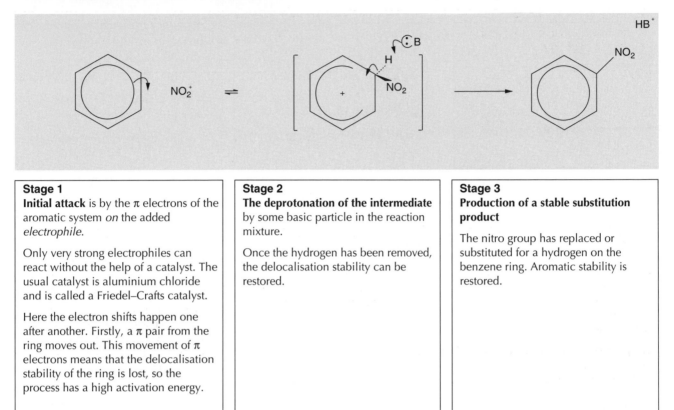

Stage 1
Initial attack is by the π electrons of the aromatic system *on* the added *electrophile*.

Only very strong electrophiles can react without the help of a catalyst. The usual catalyst is aluminium chloride and is called a Friedel–Crafts catalyst.

Here the electron shifts happen one after another. Firstly, a π pair from the ring moves out. This movement of π electrons means that the delocalisation stability of the ring is lost, so the process has a high activation energy.

Stage 2
The deprotonation of the intermediate by some basic particle in the reaction mixture.

Once the hydrogen has been removed, the delocalisation stability can be restored.

Stage 3
Production of a stable substitution product

The nitro group has replaced or substituted for a hydrogen on the benzene ring. Aromatic stability is restored.

Heterolytic electrophilic addition:
alkenes with halogens and hydrogen halides

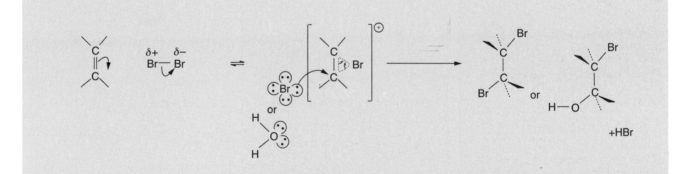

Stage 1
Inititial attack is by π electrons of an *unsaturated system* on an *electrophile*.

Sometimes the electrophile is *induced* or caused by the π system.

Notice here that, unlike the mechanism above, the electron shifts happen at the same time. As the π pair moves towards the electrophile a σ pair moves onto the far end of the bond in the electrophile. This bond is breaking *heterolytically*.

Stage 2
Nucleophilic attack on the intermediate. The intermediate is attacked from the *other side* by the *nucleophile* that was made in the *heterolysis*.

π electrons are less strongly held, so the activation energy of the first step is low and the reaction goes fast in the cold.

If the reaction is done in water, then the second attack might be by a water molecule, which has nucleophilic properties.

Stage 3
Production of *trans*-addition product

The result of the whole reaction is addition across the double bond.

Because the second attack is from the other side, the addition is always *trans* so that one part is joined on one side of the chain and the other on the other side of the chain.

If the substance being added is HX, then the H goes onto the carbon with the most hydrogen atoms already bonded to it (Markovnikov's Rule).

Homolytic substitution:
a free radical photochemical chain reaction

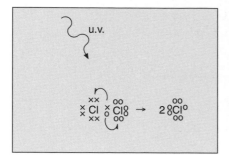

$$Cl \cdot \quad H—CH_3 \quad \rightarrow \quad HCl + CH_3\cdot$$

$$CH_3\cdot \quad Cl—Cl \quad \rightarrow \quad HCl + Cl\cdot$$

$$Cl\cdot + H—CH_2Cl \rightarrow HCl + CH_2Cl\cdot$$

$$CH_2Cl\cdot + Cl—Cl \rightarrow CH_2Cl_2 + Cl\cdot$$

$$CH_3\cdot \quad \cdot Cl \quad \rightarrow \quad CH_3 Cl$$

Stage 1
Initiation: the production of **free radicals**.

A chlorine molecule absorbs ultraviolet radiation. This energy makes the bond in the molecule break *homolytically* producing two *free radicals*.

Because this reaction is started by light, it is a **photochemical** reaction.

Stage 2
Propagation: the reaction between a *free radical* and a molecule to make another *free radical*.

There are many possible steps.

Because each step makes another reactive free radical, the reaction is a **chain reaction**.

Stage 3
Termination: the reaction between *two free radicals* making an *unreactive molecule*.

Because any two free radicals can meet, this type of reaction always makes a mixture of products.

Homolytic addition:
a free radical polymerisation reaction

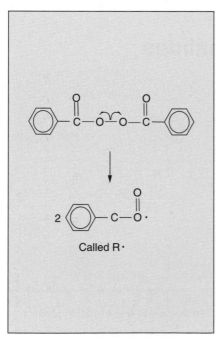

Called R ·

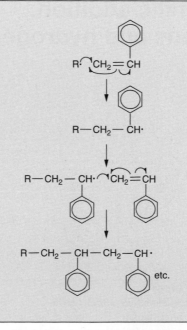

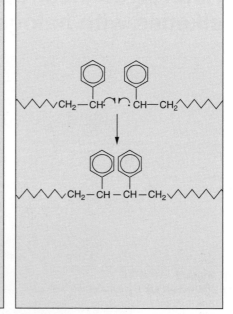

Stage 1
Initiation: the production of *free radicals*.

An initiator, usually an organic peroxide, is heated. This energy makes the oxygen–oxygen bond in the molecule break *homolytically* producing two *free radicals*.

Stage 2
Propagation: the reaction between a *growing chain*, which is a free radical, and a *monomer* molecule to make a longer growing chain radical.

There are many possible steps.

Because each step makes another reactive free radical, the reaction is a chain reaction.

Stage 3
Termination: *two growing chains* meet leading to the disappearance of the radical active sites.

Because any two free radicals can meet, this type of reaction always makes a mixture of chains of different lengths.

Reaction pathways

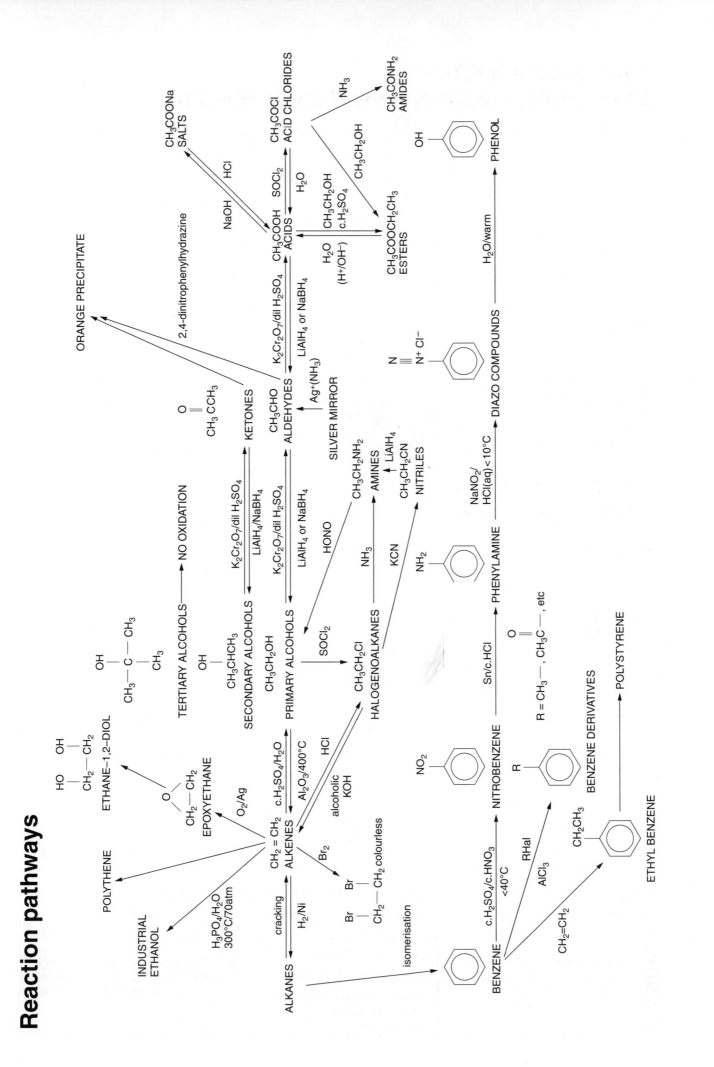

Alkanes: the facts

SATURATED HYDROCARBONS

The alkanes:
- all have the same general formula C_nH_{2n+2}
- all have only single bonds in the chain
- show a gradation of physical properties: as the chain increases in length the fixed points and viscosity increase and the colour darkens
- all have similar chemical reactions

A group of compounds with these properties is an example of a **homologous series.**

SOURCE OF ALKANES

Alkanes are found in **crude oil or petroleum** which is the remains of huge numbers of very small marine animals. Crude oil is a *mixture* of alkanes; the composition of this mixture changes from one source to the next. Natural gas is a source of methane.

fractions	Arabian heavy crude	North Sea crude
gas/petrol	18%	23%
jet fuel	11%	15%
heating oil	18%	24%
fuel oil	53%	38%

SEPARATION OF CRUDE OIL

Crude oil is a mixture of substances with boiling points that are close together. They are physically separated using *fractional distillation*. Each *fraction* is a group of saturated hydrocarbons whose boiling points are very close together.

Uses of crude oil fractions

The fractions of crude oil have three main kinds of use:

1. as *fuels* because they react exothermically with air producing non-toxic products;

2. *for their physical properties*: as lubricants, greases, tar on roads and roofs;

3. *as a source of petrochemicals*: decomposition and rearrangement can make other useful compounds.

CRACKING

The supply of each fraction does not always match the demand. Larger volumes of petrol and diesel are needed than are produced. If more crude oil is fractionally distilled to meet this demand there will be too much of other fractions. This **supply/demand mismatch** is avoided by decomposing some of the higher boiling point fractions. Their large molecules can be broken down into smaller ones for use in petrol. This process of *decomposition* is called **cracking.** Cracking also makes *unsaturated hydrocarbons* which can be made into ethanol or polythene.

$$C_{17}H_{36} \xrightarrow[400\ ^{\circ}C]{Al_2O_3} C_8H_{18} + CH_3CH=CH_2 + CH_2=CH_2$$

petrol → polypropylene → polythene and methylated spirits

ALKANE REACTIONS

1. **Burning** Alkanes burn in air to make carbon dioxide and water.

$$C_9H_{20}(l) + 14O_2(g) \rightarrow 9CO_2(g) + 10H_2O(g)$$

But in the restricted amount of air in the cylinder of a petrol engine carbon monoxide is also made. The energy of the spark can also make the nitrogen of the air react. The oxides of nitrogen and the carbon monoxide are both toxic. They can be removed from the exhaust gases in a catalytic converter. Any lead in the petrol will poison the catalyst so unleaded petrol must be used.

$$2CO(g) + 2NO(g) \xrightarrow{Pt} 2CO_2(g) + N_2(g)$$

2. **With chlorine** Alkanes react rapidly and violently with chlorine or bromine in the presence of ultraviolet radiation. A mixture of products is always made.

$$CH_4 + Cl_2 \rightarrow CH_3Cl + HCl$$
$$CH_3Cl + Cl_2 \rightarrow CH_2Cl_2 + HCl$$
$$CH_2Cl_2 + Cl_2 \rightarrow CHCl_3 + HCl$$
$$CHCl_3 + Cl_2 \rightarrow CCl_4 + HCl$$

Bromine reacts in a similar way.

Alkanes: the theory

GRADATION OF PHYSICAL PROPERTIES

Each alkane has strong covalent bonds inside the molecule. There is very little charge separation along these bonds so there are no very reactive sites.

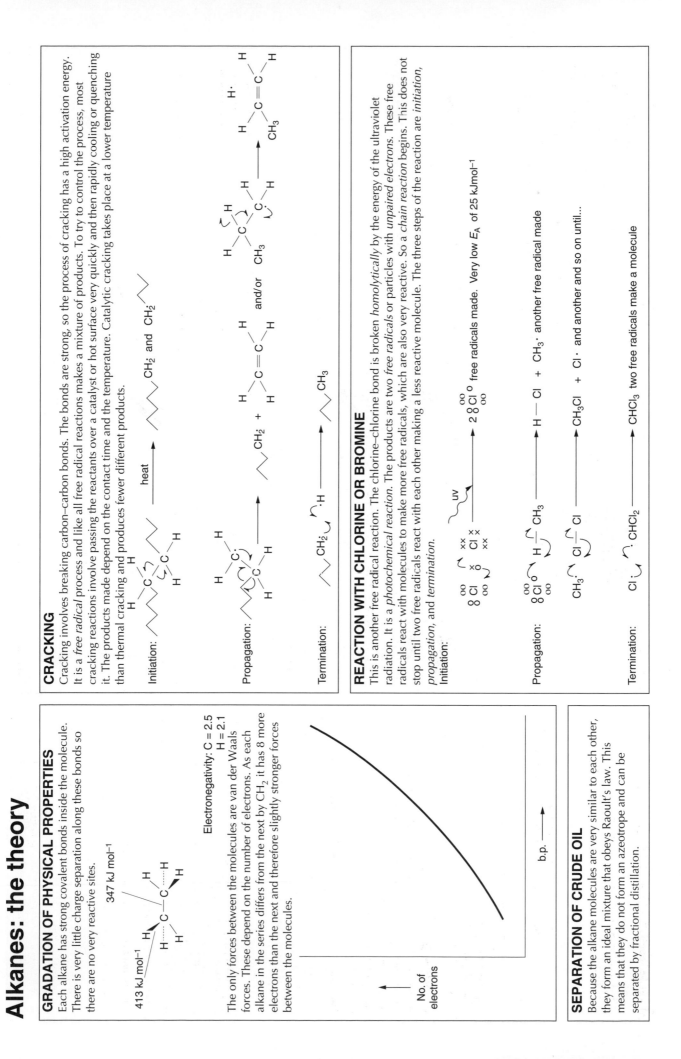

413 kJ mol^{-1}

347 kJ mol^{-1}

Electronegativity: C = 2.5
H = 2.1

The only forces between the molecules are van der Waals forces. These depend on the number of electrons. As each alkane in the series differs from the next by CH_2 it has 8 more electrons than the next and therefore slightly stronger forces between the molecules.

No. of electrons

b.p.

SEPARATION OF CRUDE OIL

Because the alkane molecules are very similar to each other, they form an ideal mixture that obeys Raoult's law. This means that they do not form an azeotrope and can be separated by fractional distillation.

CRACKING

Cracking involves breaking carbon–carbon bonds. The bonds are strong, so the process of cracking has a high activation energy. It is a *free radical* process and like all free radical reactions makes a mixture of products. To try to control the process, most cracking reactions involve passing the reactants over a catalyst or hot surface very quickly and then rapidly cooling or quenching it. The products made depend on the contact time and the temperature. Catalytic cracking takes place at a lower temperature than thermal cracking and produces fewer different products.

Initiation:

Propagation:

Termination:

REACTION WITH CHLORINE OR BROMINE

This is another free radical reaction. The chlorine–chlorine bond is broken *homolytically* by the energy of the ultraviolet radiation. It is a *photochemical reaction*. The products are two *free radicals* or particles with *unpaired electrons*. These free radicals react with molecules to make more free radicals, which are also very reactive. So a *chain reaction* begins. This does not stop until two free radicals react with each other making a less reactive molecule. The three steps of the reaction are *initiation, propagation,* and *termination.*

Initiation:

$2 \, \overset{\circ\circ}{\underset{\circ\circ}{Cl}} \, \circ$ free radicals made. Very low E_A of 25 kJmol⁻¹

Propagation:

$H-Cl + CH_3 \cdot$ another free radical made

$CH_3Cl + Cl \cdot$ and another and so on until...

Termination:

$CHCl_3$ two free radicals make a molecule

Alkenes: the facts

Addition reactions

1. With halogens

The reaction with bromine is an important diagnostic test because the bromine is decolorised as it adds on.

$$H_2C = CH_2 + Br_2$$

brown → colourless

2. With hydrogen compounds

(a) hydrogen halides

$$+ HBr$$

(b) concentrated sulphuric acid

$$+ HHSO_4$$

if water is now added H_2O

$$CH_3CH_2OH + H_2SO_4$$

3. With water

Industrially ethene is hydrated to ethanol, used as a solvent and fuel – methylated spirits

$$CH_2 = CH_2 + H_2O \xrightarrow[300°C + 70\ atm.]{H_3PO_4,\ water} CH_3CH_2OH$$

4. With oxygen

This is another important industrial process. **Epoxyethane** will undergo hydrolysis to give ethane-1,2–diol for antifreeze and polyesters.

$$2CH_2 = CH_2 + O_2 \xrightarrow[250°C]{Ag} 2\ CH_2 - CH_2\ epoxyethane$$

then

$$CH_2 - CH_2 + H_2O \xrightarrow{200°C} CH_2 - CH_2\ ethane-1,2-diol$$

5. With acidified manganate (VII) solution.

$$CH_2 = CH_2 + H_2O \xrightarrow[dil\ H_2SO_4]{KMnO_4} CH_2 - CH_2\ ethane-1,2-diol$$

6. With itself making a polymer

$$n(CH_2 = CH_2) \longrightarrow (CH_2 - CH_2)_n$$

ethene monomer → polythene

POLYMERISATION

Because they are unsaturated, alkene molecules can add onto each other forming longer chains. When this process is allowed to go on for some time a very much longer molecule called a polymer is formed.

ANIMAL FATS, VEGETABLE OILS, AND HEART DISEASE.

Both animal fats and vegetable oils are long chain carboxylic acids. Fats are solid at room temperature while oils are liquid. The important difference between them is that the carbon chains in the animal fat are saturated while the chains in vegetable oils are unsaturated and have double bonds in them.

in a fat CH_3 ... acid chain saturated

in an oil CH_3 ... acid chain unsaturated

There appears to be a correlation between the amount of animal fat in human diet and the incidence of heart disease. Populations which consume high fat diets have more heart disease than those who eat and cook mainly with vegetable oils. This has resulted in marketing campaigns for *polyunsaturated fats*.

Polyunsaturated fatty acids are important for healthy metabolism

e.g. γ linolenic acid CH_3 ...

and arachidonic acid CH_3 ...

This is a contradiction because to make a solid margarine, the manufacturer has *hardened* it by reacting it with hydrogen over a nickel catalyst. Recently, evidence suggests that many of these margarines are no better for our health than natural fats.

$$+ H_2 \xrightarrow{Ni}$$

an oil → saturated chain like a fat: a hardened oil

Alkenes: the theory

STABILITY AND REACTIVITY

The bonding between the two doubly bonded carbon atoms is stronger than the single bond in alkanes.

$$\diagdown C = C \diagup \quad 612 \text{ kJ mol}^{-1} : \text{ not twice as much!}$$

$$\diagdown C - C \diagup \quad 348 \text{ kJ mol}^{-1}$$

This makes the molecules more thermally stable than alkanes.

The strength of the double bond is not twice that of a single bond, which reminds us that it is a different kind of bond.

σ overlap

π overlap

π bond — two regions of negative charge are a reactive site

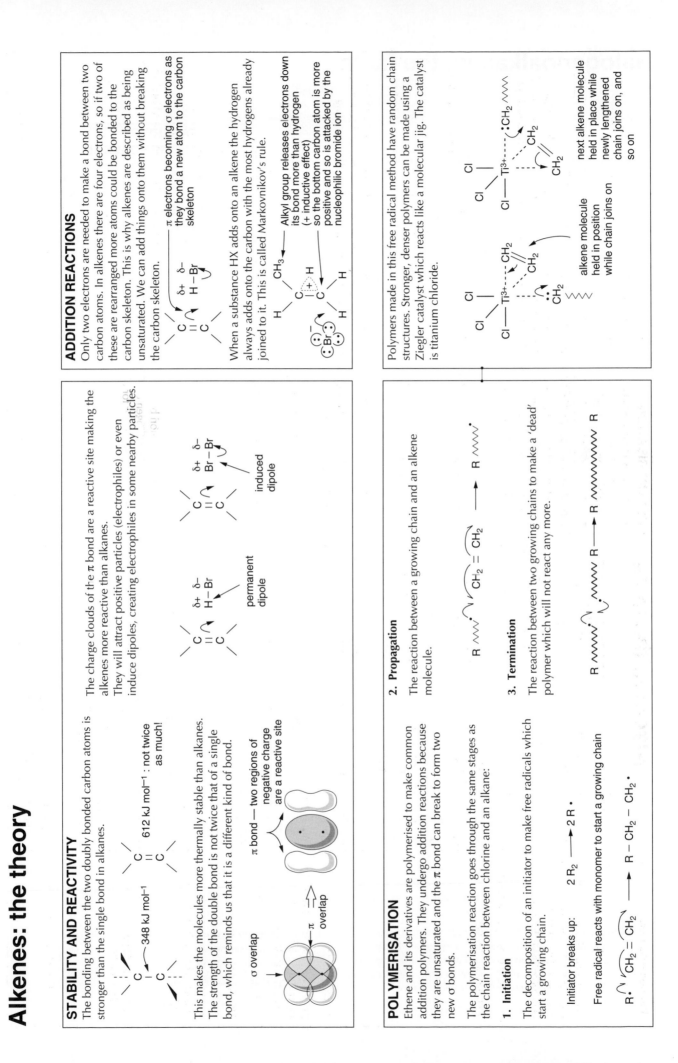

ADDITION REACTIONS

Only two electrons are needed to make a bond between two carbon atoms. In alkenes there are four electrons, so if two of these are rearranged more atoms could be bonded to the carbon skeleton. This is why alkenes are described as being unsaturated. We can add things onto them without breaking the carbon skeleton.

π electrons becoming σ electrons as they bond a new atom to the carbon skeleton

$$\diagdown C = C \diagup \quad \overset{\delta+ \quad \delta-}{H - Br}$$

When a substance HX adds onto an alkene the hydrogen always adds onto the carbon with the most hydrogens already joined to it. This is called Markovnikov's rule.

Alkyl group releases electrons down its bond more than hydrogen (+ inductive effect) so the bottom carbon atom is more positive and so is attacked by the nucleophilic bromide ion

The charge clouds of the π bond are a reactive site making the alkenes more reactive than alkanes.

They will attract positive particles (electrophiles) or even induce dipoles, creating electrophiles in some nearby particles.

$$\diagdown C = C \diagup \quad \overset{\delta+ \quad \delta-}{H - Br} \quad \text{permanent dipole}$$

$$\diagdown C = C \diagup \quad \overset{\delta+ \quad \delta-}{Br - Br} \quad \text{induced dipole}$$

POLYMERISATION

Ethene and its derivatives are polymerised to make common addition polymers. They undergo addition reactions because they are unsaturated and the π bond can break to form two new σ bonds.

The polymerisation reaction goes through the same stages as the chain reaction between chlorine and an alkane:

1. Initiation

The decomposition of an initiator to make free radicals which start a growing chain.

Initiator breaks up: $2 R_2 \longrightarrow 2 R\cdot$

Free radical reacts with monomer to start a growing chain

$$R\cdot \quad CH_2 = CH_2 \longrightarrow R - CH_2 - CH_2\cdot$$

2. Propagation

The reaction between a growing chain and an alkene molecule.

$$R \wedge\wedge\wedge\cdot \quad CH_2 = CH_2 \longrightarrow R \wedge\wedge\wedge\cdot$$

3. Termination

The reaction between two growing chains to make a 'dead' polymer which will not react any more.

$$R \wedge\wedge\wedge\cdot \quad \cdot\wedge\wedge\wedge R \longrightarrow R \wedge\wedge\wedge\wedge\wedge\wedge R$$

Polymers made in this free radical method have random chain structures. Stronger, denser polymers can be made using a Ziegler catalyst which reacts like a molecular jig. The catalyst is titanium chloride.

alkene molecule held in position while chain joins on

next alkene molecule held in place while newly lengthened chain joins on, and so on

Halogenoalkanes: the facts

Halogenoalkanes, also known as haloalkanes or alkyl halides:

CH_3CH_2Cl $BrCH_2CH_2Br$ CHI_3

CLASSES OF HALOGENOALKANE
Like alcohols they can be classified depending on the number of carbon atoms joined to the carbon next to the halogen atom.

Primary: one carbon joined

Secondary: two carbons joined

Tertiary: three carbons joined

POLLUTION PROBLEMS WITH HALOGENOALKANES
The widespread use of halogenoalkanes leads to pollution problems:

1. CFCs which have escaped from aerosols and fridges are decomposed in the upper atmosphere and then react with ozone (see p. 94)

2. Aromatic halogen compounds and alkanes in which three or four halogen atoms are joined to the same carbon atom are not hydrolysed — they are not biodegradable — but are soluble in body fats. This means that they get concentrated in organisms, which are then eaten by predators. This results in an ever increasing concentration of these compounds in the body tissues of animals further up the food chain. This has been blamed for falling fertility rates and other threats to health.

REACTIONS
The first three reactions below involve nucleophilic reagents

1. **With the hydroxide ion**: the hydroxide ion can take the place of the halogen atom in a substitution reaction; the reaction is slow and done under reflux.

$$CH_3CH_2Cl(l) + OH^-(aq) \xrightarrow{reflux} CH_3CH_2OH(aq) + Cl^-(aq)$$

The extent of this reaction can be followed by adding silver nitrate solution. Silver ions will react with free halide ions to form silver halide precipitate, but not with the covalently bonded halogen atoms. So, as the halogen atoms are substituted and leave as halide ions, they react to form a visible precipitate.

2. **With the cyanide ion**: the cyanide ion can also take the place of the halogen; this reaction is important because it makes a carbon–carbon bond and increases the length of the carbon chain. This reaction is also slow, and is carried out in alcohol with a trace of alkali under reflux.

$$CH_3CH_2Br(l) + CN^-(aq) \xrightarrow{reflux} CH_3CH_2CN(l) + Br^-(aq)$$

3. **With ammonia**: concentrated ammonia ('880 ammonia') reacts with halogenoalkanes to make amines. In this reaction each organic substance made is a more reactive nucleophile than the ammonia. So the new substance will attack another halogenoalkane molecule making an even more reactive nucleophile than the previous one, and so on. Once a reaction starts it goes all the way to the quaternary ammonium compound.

$$NH_3(g) + CH_3CH_2Cl(l) \rightarrow CH_3CH_2NH_2(l) + HCl(g)$$
$$CH_3CH_2NH_2(l) + CH_3CH_2Cl(l) \rightarrow (CH_3CH_2)_2NH(l) + HCl(g)$$
$$(CH_3CH_2)_2NH(l) + CH_3CH_2Cl(l) \rightarrow (CH_3CH_2)_3N(l) + HCl(g)$$
$$(CH_3CH_2)_3N(l) + CH_3CH_2Cl(l) \rightarrow (CH_3CH_2)_4N^+ + Cl^-(l)$$

4. **With alcoholic hydroxide ions**: here the hydroxide ion reacts as a base instead of a nucleophile and so an elimination reaction takes place:

$$CH_3CH_2CHBr(l) + OH^-(alc)$$
$$\rightarrow CH_3CH = CH_2(g) + H_2O(l) + Br^-(alc)$$

Halogenoalkanes: the theory

REACTIVE SITE

	halogen	electronegativity
Ⓒ 2.5	Ⓕ	4.0
	Ⓒˡ	3.0
	Ⓑʳ	2.8
	Ⓘ	2.5

Most halogens are more electronegative than carbon. Halogen atoms bonded to carbon atoms therefore pull electron density away from the carbon atom making it electron deficient. The covalent bond is polarised as the charge along it is separated. This polarisation of a covalent bond is known as the **inductive effect**.

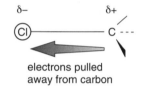

electrons pulled
away from carbon

Nucleophiles will attack the carbon atom.

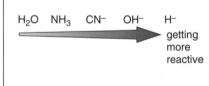

H_2O NH_3 CN^- OH^- H^-

getting more reactive

ATTACK LEADS TO AN S_N REACTION

This stands for *substitution* by a *nucleophile*. If the halide is primary, then the mechanism is S_N2 and involves a collision between the incoming nucleophile and the electron deficient carbon atom. If the halide is a tertiary one, the nucleophile's approach is sterically hindered, but the carbon–halogen bond can break more easily forming a reactive carbocation. Once this forms it reacts very quickly with the nucleophile. As the formation of the carbocation is the rate determining step the mechanism is S_N1.

for primary halides the rate determining step is this collision

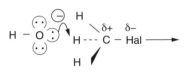

for tertiary halides the rate determining step is the formation of the carbocation

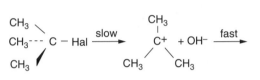

ELIMINATION

This can sometimes happen instead of substitution. This is most likely with tertiary halides because:

1. the alkyl groups get in the way of the incoming nucleophile and

2. the alkyl groups release electrons towards the reactive carbon making it a less reactive site.

The attacking particle acts as a base, removing a hydrogen from the edge of the molecule, because its nucleophilic pathway is blocked.

Substitution and elimination are always in competition. Substitution is most likely with a primary halide and elimination with a tertiary halide.

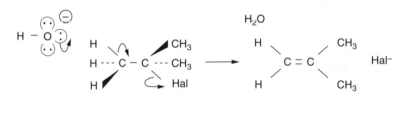

The reactivity of the halogen compound depends on the type of carbon skeleton and on the halogen.

For substitution reactions with a given halide the order of reactivity is:

primary halide < secondary halide < tertiary halide.

If the halogen atoms change, the order of reactivity is dictated by the strength of the bond between the carbon and the halogen atom and so is:

$RCl < RBr < RI$

If the carbon skeleton changes from alkyl to aryl to acyl, this affects reactivity too:

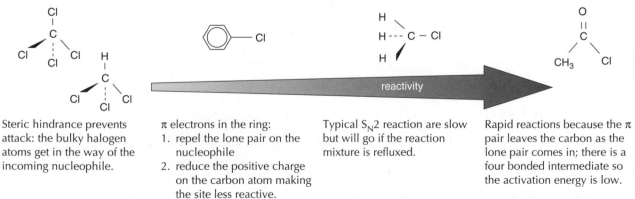

reactivity

Steric hindrance prevents attack: the bulky halogen atoms get in the way of the incoming nucleophile.

π electrons in the ring:
1. repel the lone pair on the nucleophile
2. reduce the positive charge on the carbon atom making the site less reactive.

Typical S_N2 reaction are slow but will go if the reaction mixture is refluxed.

Rapid reactions because the π pair leaves the carbon as the lone pair comes in; there is a four bonded intermediate so the activation energy is low.

Alcohols: the facts

Alcohols are derived from water: they are *alkylated* water

$H - O \diagdown_H$ replace one H by a carbon chain $CH_3 - O \diagdown_H$ or $CH_3CH_2 - O \diagdown_H$

There is hydrogen bonding between the molecules and between alcohol molecules and water. This means that alcohols are miscible (soluble) in water.

CLASSES OF ALCOHOLS

Alcohols are classed depending on the number of carbons joined to carbon next to the OH group

Primary: one carbon joined   **Secondary:** two carbons joined   **Tertiary:** three carbons joined

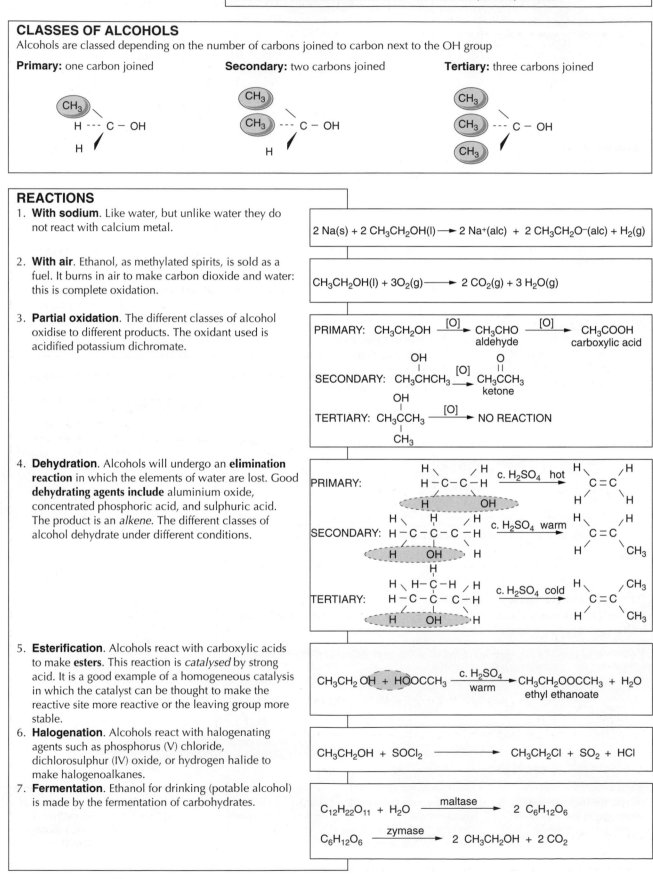

REACTIONS

1. **With sodium**. Like water, but unlike water they do not react with calcium metal.

$$2\,Na(s) + 2\,CH_3CH_2OH(l) \longrightarrow 2\,Na^+(alc) + 2\,CH_3CH_2O^-(alc) + H_2(g)$$

2. **With air**. Ethanol, as methylated spirits, is sold as a fuel. It burns in air to make carbon dioxide and water: this is complete oxidation.

$$CH_3CH_2OH(l) + 3O_2(g) \longrightarrow 2\,CO_2(g) + 3\,H_2O(g)$$

3. **Partial oxidation**. The different classes of alcohol oxidise to different products. The oxidant used is acidified potassium dichromate.

PRIMARY: $CH_3CH_2OH \xrightarrow{[O]} CH_3CHO \xrightarrow{[O]} CH_3COOH$
     aldehyde   carboxylic acid

SECONDARY: $CH_3\underset{OH}{\overset{OH}{CH}}CH_3 \xrightarrow{[O]} CH_3\underset{O}{\overset{O}{C}}CH_3$ ketone

TERTIARY: $CH_3\underset{CH_3}{\overset{OH}{C}}CH_3 \xrightarrow{[O]}$ NO REACTION

4. **Dehydration**. Alcohols will undergo an **elimination reaction** in which the elements of water are lost. Good **dehydrating agents include** aluminium oxide, concentrated phosphoric acid, and sulphuric acid. The product is an *alkene*. The different classes of alcohol dehydrate under different conditions.

PRIMARY: $\xrightarrow{\text{c. } H_2SO_4 \text{ hot}}$

SECONDARY: $\xrightarrow{\text{c. } H_2SO_4 \text{ warm}}$

TERTIARY: $\xrightarrow{\text{c. } H_2SO_4 \text{ cold}}$

5. **Esterification**. Alcohols react with carboxylic acids to make **esters**. This reaction is *catalysed* by strong acid. It is a good example of a homogeneous catalysis in which the catalyst can be thought to make the reactive site more reactive or the leaving group more stable.

$$CH_3CH_2OH + HOOCCH_3 \xrightarrow[\text{warm}]{\text{c. } H_2SO_4} CH_3CH_2OOCCH_3 + H_2O$$
     ethyl ethanoate

6. **Halogenation**. Alcohols react with halogenating agents such as phosphorus (V) chloride, dichlorosulphur (IV) oxide, or hydrogen halide to make halogenoalkanes.

$$CH_3CH_2OH + SOCl_2 \longrightarrow CH_3CH_2Cl + SO_2 + HCl$$

7. **Fermentation**. Ethanol for drinking (potable alcohol) is made by the fermentation of carbohydrates.

$$C_{12}H_{22}O_{11} + H_2O \xrightarrow{\text{maltase}} 2\,C_6H_{12}O_6$$

$$C_6H_{12}O_6 \xrightarrow{\text{zymase}} 2\,CH_3CH_2OH + 2\,CO_2$$

Alcohols: the theory

FUNCTIONAL GROUP AND REACTIVE SITES

The functional group is the hydroxy group. It is the oxygen atom of the group, with its lone pairs and high electronegativity, which makes alcohols reactive.

electron deficient carbon attacked by nucleophiles

some basic properties like water

H δ+ some acidic properties like water

REDUCTION WITH SODIUM

The hydrogen in the hydroxy group is electron deficient like the hydrogen in a water molecule. This means that it will attack the delocalised electrons in sodium. However, the alkyl group is electron releasing compared to hydrogen and this makes the alcohol hydrogen less positive than the water one, so while water can react with the more tightly held delocalised electrons in calcium, alcohols will not.

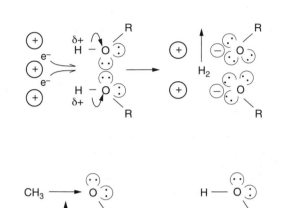

CH_3 → O with H δ+   H — O with H δ+

alkyl group releases electrons compared to hydrogen

ESTERIFICATION

The lone pair on the hydroxy group attacks the acid nucleophilically. This attack is catalysed by the presence of strong acid. The acid catalyst can be thought of as enhancing the reactive site and making a better leaving group.

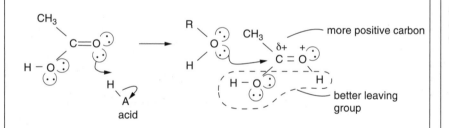

more positive carbon

better leaving group

acid

PARTIAL OXIDATION

The process of partial oxidation involves the hydrogens on the carbon joined to the hydroxy group.

There are two of these on a primary alcohol so it is oxidised to an aldehyde and then acid.

$CH_3 - C - OH$ two hydrogens $\Rightarrow$ two step oxidation ⟶ aldehyde ⟶ acid

There is only one on a secondary alcohol so it is oxidised just to a ketone.

$CH_3 - C - OH$ one hydrogen $\Rightarrow$ one step oxidation ⟶ ketone
CH_3

There are none on a tertiary alcohol so it is not partially oxidised at all.

CH_3
$CH_3 - C - OH$ no hydrogens $\Rightarrow$ no oxidation
CH_3

DEHYDRATION

The strong acid first protonates the alcohol making a better leaving group and then this *carbocation* is deprotonated and water is lost.

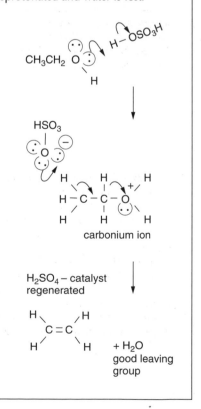

CH_3CH_2 O H — OSO_3H

HSO_3

carbonium ion

H_2SO_4 – catalyst regenerated

C = C + H_2O good leaving group

Carbonyl compounds: the facts

REACTIONS

1. Redox

Aldehydes can be reduced to primary alcohols or oxidised to acids.

Ketones can only be reduced to secondary alcohols.

In both cases the reducing agent is either lithium aluminium hydride or sodium borohydride. The oxidising agent is acidified potassium dichromate.

The fact that aldehydes can be oxidised means that they are *reducing agents*. This fact is the basis of two *diagnostic tests* used to distinguish between aldehydes and ketones. Aldehydes will reduce silver ions in ammonia solution (Tollen's reagent) to silver metal, and copper (II) ions in a special solution (either Fehling's or Benedict's solution) to copper (I).

$$CH_3CH_2OH \xrightarrow[\text{reduction}]{[H]} CH_3CHO \xrightarrow[\text{oxidation}]{[O]} CH_3COOH$$
primary alcohol   aldehyde   acid

$$CH_3CHCH_3 \xrightarrow[\text{reduction}]{[H]} CH_3CCH_3$$
$\quad$ OH $\qquad\qquad\qquad$ O
secondary alcohol $\qquad$ ketone

$$CH_3CHO + Ag^+(NH_3) \longrightarrow CH_3COOH + Ag(s)$$ Grey black ppt. or silver mirror seen
aldehyde

$$CH_3CHO + Cu^{2+}(aq) \longrightarrow CH_3COOH + Cu^+$$ Yellowish ppt. seen
aldehyde

2. With hydrogen cyanide

Both aldehydes and ketones react with hydrogen cyanide in the presence of a catalyst of potassium cyanide.

$$CH_3CHO + HCN \xrightarrow{KCN} CH_3-\underset{H}{\overset{CN}{\underset{|}{\overset{|}{C}}}}-OH$$

3. With 2,4-dinitrophenylhydrazine

Both aldehydes and ketones react with this reagent giving an orange precipitate. This is a *diagnostic test for the carbonyl group.*

$$CH_3CHO \xrightarrow{\text{2,4-DNPH}} \text{orange ppt.}$$

$$CH_3CCH_3 \xrightarrow{\text{2,4-DNPH}} \text{orange ppt.}$$
$\;$ O (double bond)

4. The iodoform test

When the carbonyl group is one carbon in from the end of the carbon chain it will react with iodine in alkaline solution making a yellow solid, CHI_3, tri-iodomethane, previously called iodoform. Because secondary alcohols can be oxidised by iodine, this reaction is also done by secondary alcohols in which the OH group is joined to the second carbon.

$$CH_3CR \xrightarrow{\text{I}_2 \text{ in OH}^-(aq)} RCOO^-(aq) + CHI_3(s) \longrightarrow \text{yellow ppt.}$$
$\;$ O (double bond)

also done by $CH_3CH(OH)R$

So the iodoform test is a diagnostic test for the CH_3CO- and the $CH_3CH(OH)-$ groups

ALDEHYDES AND KETONES

If the carbonyl group is at the end of a carbon chain the substance is an **aldehyde**; if the carbonyl is in the chain, the substance is a **ketone**.

$$CH_3CH_2C{\overset{O}{\underset{}{\parallel}}}H$$
carbonyl at the end of the chain

aldehyde

$$CH_3CCH_3$$
$\;$ O (double bond)
carbonyl in the chain

ketone

Carbonyl compounds: the theory

π pair becomes a lone pair

FUNCTIONAL GROUP AND REACTIVE SITES

Carbonyl compounds react rapidly because the *electron deficient carbon* in the reactive site is *unsaturated*, and so as the electron pair from a nucleophile approaches, the π pair can easily move onto the oxygen atom.

Like alcohols, the aldehyde can be oxidised because it has a hydrogen joined to the carbon.

NUCLEOPHILIC ADDITION WITH HYDROGEN CYANIDE

This is one of the key mechanisms you have to learn. The potassium cyanide is a good example of a *heterogeneous catalyst*. It works by providing a better attacking group, which is the cyanide ion (hydrogen cyanide has hardly any nucleophilic properties because the nitrogen is electronegative and pulls its lone pair in). Another cyanide ion is regenerated later in the reaction replacing the one which did the initial attack.

catalyst regenerated

Net result is the addition of HCN across the double bond

DIAGNOSTIC TEST WITH 2,4-DINITROPHENYLHYDRAZINE

This reaction is an example of an addition–elimination. Once the initial attack has happened, an electron rearrangement leads to the elimination of a water molecule. The dinitrophenylhydrazone which forms is a bright coloured, insoluble product and so this is a good diagnostic test.

proton rearrangement and elimination of water

bright orange insoluble product

Carboxylic acids: the facts

REACTIONS

1. With water

The carboxylic acids are weak acids. A 1 mol dm^{-3} solution of ethanoic acid is less than 0.1 % split up into ions.

$$CH_3COOH(aq) + H_2O(l) \rightleftharpoons CH_3COO^-(aq) \qquad K_a = 1.8 \times 10^{-5} \text{ mol dm}^{-3}$$
99.9 %   <0.1 %

2. With bases

(a) Metal oxides

$$CuO(s) + 2CH_3COOH(aq) \rightarrow Cu^{2+}(aq) + 2CH_3COO^-(aq) + H_2O(l)$$

(b) Metal hydroxides

$$Na^+(aq) + OH^-(aq) + CH_3COOH(aq) \rightarrow CH_3COO^-(aq) + Na^+(aq) + H_2O(l)$$

(c) Metal carbonates

$$MgCO_3(s) + 2CH_3COOH(aq) \rightarrow Mg^{2+}(aq) + 2CH_3COO^-(aq) + H_2O(l) + CO_2(g)$$

(d) With ammonia

$$NH_3(aq) + CH_3COOH(aq) \rightarrow CH_3COO^-(aq) + NH_4^+(aq)$$

3. With alcohols

Acids react very slowly with alcohols to make a carboxylic acid ester. This reaction can be catalysed by adding some strong acid, but even so the reaction is reversible and so the yield will not be good.

$$CH_3COOH(l) + HOCH_2CH_3(l) \overset{H^+}{\rightleftharpoons} CH_3COOCH_2CH_3(l) + H_2O(l)$$
$K_c = 4$ Yield 67 %

4. With lithium aluminium hydride or sodium borohydride

Acids can be reduced to alcohols.

$$CH_3COOH(l) \xrightarrow[\text{(ii) } H_2O]{\text{(i) LiAlH}_4} CH_3CH_2OH(l)$$

5. With dichlorosulphur (IV) oxide or phosphorus (V) chloride

Acids can be halogenated

$$CH_3COOH(l) + SOCl_2(l) \rightarrow CH_3COCl(l) + SO_2(g) + HCl(g)$$

$$RCOOH(l) \xrightarrow{PCl_5} RCOCl(l)$$

Carboxylic acids: the theory

A COMBINATION OF TWO FUNCTIONAL GROUPS

The carboxylic acid functional group is a combination of the carbonyl and alcohol groups. We can still see some of the properties of these two groups in the reactions of acids. They react with lithium aluminium hydride like carbonyls:

and they react with dichlorosulphur(IV) oxide like alcohols:

But the groups are so close together that they modify each other producing new properties – the acidity.

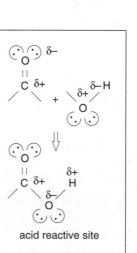

acid reactive site

MECHANISM OF ESTERIFICATION

The attacking group in this reaction is the alcohol which is not a very strong nucleophile. The positive charge on the carbonyl carbon of the acid group is reduced compared to that in aldehydes and ketones by the lone pairs on the nearby oxygen. Adding acid protonates the oxygen increasing the positive charge and making a better leaving group.

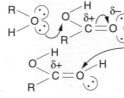

is a slow reaction, but with strong acid present

this oxygen is protonated increasing the charge on the carbon and with two –OH groups on the carbon there is a better leaving group, water.

Carboxylic acid esters: the facts

MAKING ESTERS

Esters are made when acids (including inorganic acids like phosphoric acid) react with alcohols. This is not a good way to produce esters because the reaction is slow and reversible so the yield is always low.

A much better way to make esters is to acylate an alcohol with an acid chloride. The other product is hydrogen chloride gas. This leaves the reaction vessel and so no reversible reaction is possible.

Esterification:

$$CH_3COOH(l) + HOCH_2CH_3(l) \rightleftharpoons CH_3COOCH_2CH_3(l) + H_2O(l) \quad K_c = 4$$

acid + alcohol → ester + water

$$CH_3COCl(l) + HOCH_2CH_3(l) \rightarrow CH_3COOCH_2CH_3(l) + HCl(g)$$

rapid reaction; good yield

ESTER REACTIONS

With water

Esters are hydrolysed by water. The reaction is slow and catalysed by both acid and base.

$$CH_3COOCH_2CH_3(l) + H_2O(l) \overset{H^+}{\rightleftharpoons} CH_3COOH(l) + HOCH_2CH_3(l)$$

$$CH_3COOCH_2CH_3(l) + OH^-(aq) \rightarrow CH_3COO^-(aq) + HOCH_2CH_3(l)$$

With ammonia

The reaction with ammonia is similar to that with water but an amide is made.

$$CH_3COOCH_2CH_3(l) + NH_3(aq) \rightleftharpoons CH_3CONH_2(s) + HOCH_2CH_3(l)$$

USES OF ESTERS

Esters have a range of uses depending on the carbon chains joined to the ester group:

1. **Short chain esters** have fruity smells and flavours and are used in the food industry. They are also very good solvents, used in nail varnish and polystyrene cement.

2. **Esters of phthalic acid** (e.g. dibutyl phthalate) and **triesters of phosphoric acid** (e.g. trixylylphosphate) are used as plasticisers in PVC. Plasticisers lower the glass point of a polymer. 30 to 40% by mass of these high boiling point esters are added to the polymer. Plasticiser molecules get between the polymer chains, separating them, and so allowing greater movement by the chains. This makes the polymer more flexible.

3. **Long chain triesters** are naturally occurring esters which make up animal fats and vegetable oils. The carbon chains joined to the ester group are saturated in fats and unsaturated in oils. The hydrolysis of oils and fats produces propane-1,2,3-triol (glycerol) and a long chain carboxylic acid. In alkaline conditions the acid is deprotonated forming a soap.

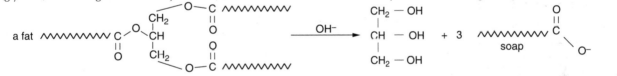

Carboxylic acid esters: the theory

HYDROLYSIS OF ESTERS

A catalyst in an organic reaction can speed the reaction up by:

1. providing a better attacking group which is regenerated later in the reaction, or

2. making the site being attacked more reactive.
   This can be done by increasing the charge on an atom or by making a better leaving group.

ALKALINE HYDROLYSIS

If alkali is added, the attacking group is the hydroxide ion instead of water. Because of its charge, it is a better nucleophile than water and so will attack faster. The leaving group deprotonates a water molecule so regenerating the catalyst.

In the presence of a lot of base, the acid which is made is deprotonated so preventing the back reaction. So as well as catalysing the reaction, base shifts the equilibrium to the right increasing the yield.

ACID HYDROLYSIS

The presence of strong acid protonates the single bonded oxygen. This makes a better leaving group and increases the positive charge on the carbonyl carbon, so making the reactive site more reactive.

protonation here makes a better leaving group

protonation here makes the carbonyl carbon more positive

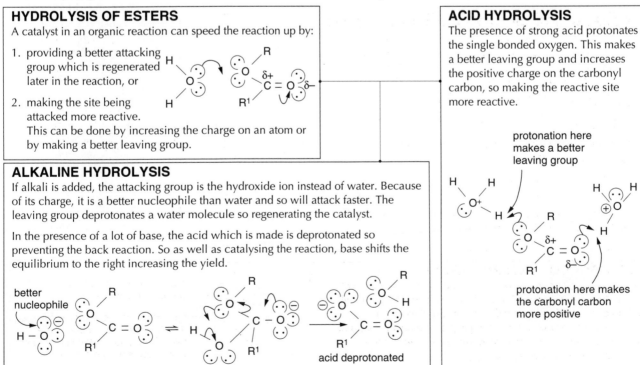

acid deprotonated

Acid chlorides and acid anhydrides: the facts

REACTIVITY
These two derivatives of carboxylic acids are very reactive. Acid chlorides are even more reactive than anhydrides and more volatile, which makes them very difficult to handle.

ACYLATION
Both react rapidly in the cold with water and ammonia and with their derivatives, alcohols and amines. In each case a hydrogen atom on the reacting molecule is replaced by an acyl group. For this reason, the reactions are all known as **acylations**, and acid chlorides and acid anhydrides are called **acylating agents**.

With acid chlorides:

water:	CH_3COCl + HOH	→ CH_3CO-OH	acid	+HCl
alcohol:	CH_3COCl + $HOCH_2CH_3$	→ $CH_3CO-OCH_2CH_3$	ester	+HCl
ammonia:	CH_3COCl + $2HNH_2$	→ CH_3CO-NH_2	amide	+NH_4Cl
amine:	CH_3COCl + $HNHCH_3$	→ $CH_3CO-NHCH_3$	substituted amide	+HCl

this hydrogen is replaced by the acyl group   HCl eliminated

With acid anhydrides:

water:	$CH_3COOOCCH_3$ + HOH	→ CH_3CO-OH	+ CH_3COOH
alcohol:	$CH_3COOOCCH_3$ + $HOCH_2CH_3$	→ $CH_3CO-OCH_2CH_3$	+ CH_3COOH
ammonia:	$CH_3COOOCCH_3$ + HNH_2	→ CH_3CO-NH_2	+ CH_3COOH
amine:	$CH_3COOOCCH_3$ + $HNHCH_3$	→ $CH_3CO-NHCH_3$	+ CH_3COOH

this hydrogen is replaced by the acyl group   ethanoic acid eliminated

Acid chlorides and acid anhydrides: the theory

The reactions above are examples of nucleophilic additions followed by eliminations.

Nucleophiles attack the carbonyl group easily because the slightly positive carbon atom has a π pair of electrons which can easily be withdrawn onto the oxygen.

The intermediate with four sigma bonds is of low energy. The activation energy of the reaction is therefore low, so it can take place quickly.

The group joined to the carbon atom (chlorine in acid chlorides, or the carboxylate group in acid anhydrides) is a good leaving group. Electron density therefore flows back towards the leaving group which is eliminated from the molecule.

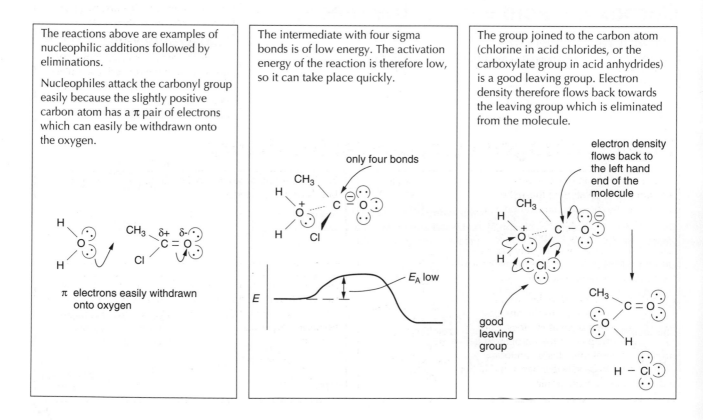

π electrons easily withdrawn onto oxygen

only four bonds

E_A low

electron density flows back to the left hand end of the molecule

good leaving group

Organic nitrogen compounds: amines, nitriles, amides, and ammonium salts

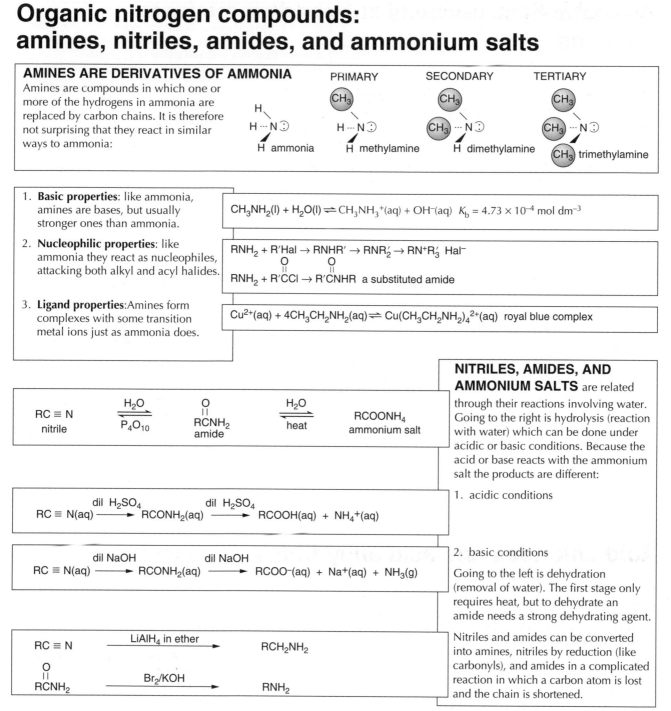

AMINES ARE DERIVATIVES OF AMMONIA

Amines are compounds in which one or more of the hydrogens in ammonia are replaced by carbon chains. It is therefore not surprising that they react in similar ways to ammonia:

PRIMARY | SECONDARY | TERTIARY

ammonia | methylamine | dimethylamine | trimethylamine

1. **Basic properties**: like ammonia, amines are bases, but usually stronger ones than ammonia.

$$CH_3NH_2(l) + H_2O(l) \rightleftharpoons CH_3NH_3^+(aq) + OH^-(aq) \quad K_b = 4.73 \times 10^{-4} \text{ mol dm}^{-3}$$

2. **Nucleophilic properties**: like ammonia they react as nucleophiles, attacking both alkyl and acyl halides.

$$RNH_2 + R'Hal \rightarrow RNHR' \rightarrow RNR_2' \rightarrow RN^+R_3' \ Hal^-$$

$$RNH_2 + R'\overset{O}{\underset{||}{C}}Cl \rightarrow R'\overset{O}{\underset{||}{C}}NHR \quad \text{a substituted amide}$$

3. **Ligand properties**: Amines form complexes with some transition metal ions just as ammonia does.

$$Cu^{2+}(aq) + 4CH_3CH_2NH_2(aq) \rightleftharpoons Cu(CH_3CH_2NH_2)_4^{2+}(aq) \quad \text{royal blue complex}$$

NITRILES, AMIDES, AND AMMONIUM SALTS

are related through their reactions involving water. Going to the right is hydrolysis (reaction with water) which can be done under acidic or basic conditions. Because the acid or base reacts with the ammonium salt the products are different:

$$RC \equiv N \underset{P_4O_{10}}{\overset{H_2O}{\rightleftharpoons}} \overset{O}{\underset{||}{R}CNH_2} \underset{\text{heat}}{\overset{H_2O}{\rightleftharpoons}} RCOONH_4$$

nitrile | amide | ammonium salt

1. acidic conditions

$$RC \equiv N(aq) \xrightarrow{\text{dil } H_2SO_4} RCONH_2(aq) \xrightarrow{\text{dil } H_2SO_4} RCOOH(aq) + NH_4^+(aq)$$

2. basic conditions

$$RC \equiv N(aq) \xrightarrow{\text{dil NaOH}} RCONH_2(aq) \xrightarrow{\text{dil NaOH}} RCOO^-(aq) + Na^+(aq) + NH_3(g)$$

Going to the left is dehydration (removal of water). The first stage only requires heat, but to dehydrate an amide needs a strong dehydrating agent.

Nitriles and amides can be converted into amines, nitriles by reduction (like carbonyls), and amides in a complicated reaction in which a carbon atom is lost and the chain is shortened.

$$RC \equiv N \xrightarrow{\text{LiAlH}_4 \text{ in ether}} RCH_2NH_2$$

$$\overset{O}{\underset{||}{R}CNH_2} \xrightarrow{Br_2/KOH} RNH_2$$

Organic nitrogen compounds: the theory

It is most helpful to think of these compounds in relation to those they resemble. Amines are like ammonia. The reactive site is the lone pair of electrons, which can attack slightly positive hydrogen atoms (acting as a base), slightly positive non-metal atoms (acting as a nucleophile), or cations (acting as a ligand).

The alkyl group joined to the nitrogen releases electrons towards the nitrogen better than a hydrogen atom does. This allows the lone pair on the nitrogen to be more prominent than it is on ammonia. So amines tend to be stronger bases and better nucleophiles than ammonia.

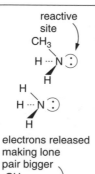

reactive site

electrons released making lone pair bigger

Nitriles are similar to carbonyl compounds in having π bonds onto the carbon.

π pairs

$$R - C \equiv N$$

Amides like acids consist of two functional groups so close together that they modify each other. In particular, the lone pair on the amide is delocalised so that amides have almost no basic properties.

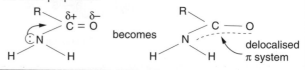

becomes

delocalised π system

Aromatic hydrocarbons: the facts

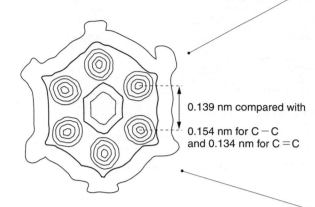

Electron density map of benzene: shows the six equal bonds in the ring

0.139 nm compared with

0.154 nm for C — C
and 0.134 nm for C = C

STRUCTURE OF BENZENE

Benzene is a symmetrical hexagonal molecule. The π bonds are delocalised around the ring and the bond lengths are between those of a C–C and a C=C.
The structure is represented by a hexagon with a circle inside it.

REACTIVITY OF BENZENE

Benzene is very unreactive.

There is no reaction with bromine or alkaline manganate (VII). The lack of reactivity of the benzene ring can be seen if alkaline manganate (VII) is added to methyl benzene. The methyl group is oxidised instead of the ring.

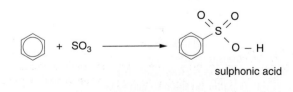

benzoic acid

1. Benzene and its derivatives can be nitrated using a mixture of concentrated nitric and sulphuric acid. The temperature must be controlled to prevent more than one nitro- group going in.

sulphuric acid protonates: $H_2SO_4 + HNO_3 \longrightarrow HSO_4 + H_2NO_3^+$

then dehydrates: $H_2NO_3^+ \xrightarrow{-H_2O} NO_2^+$ nitronium ion

the nitronium ion is a strong electrophile:

$+ NO_2^+ \xrightarrow{T < 40\ ^oC}$ NO_2 $+ H^+$

2. Benzene can be made to react with halogen derivatives using aluminium chloride as a catalyst. This is called a Friedel–Crafts reaction.

$+ R - Hal \xrightarrow{AlCl_3}$ R $+ H\ Hal$

R = Hal; or an alkyl group, e.g. CH_3; or an acyl group

$$CH_3 - \overset{\overset{\displaystyle O}{\|}}{C} -$$

3. Benzene can be sulphonated by reacting it with fuming sulphuric acid (oleum). The benzene reacts with sulphur trioxide in the oleum.

$+ SO_3 \longrightarrow$

sulphonic acid

4. Industrially benzene is converted into ethylbenzene by reacting it with ethene. The ethylbenzene (also called styrene) is used to make polystyrene.

USES OF BENZENE DERIVATIVES

Detergents

Sodium salts of sulphonic acid and its derivatives are widely used as detergents. The covalent end of the particles are attracted by van der Waals forces to organic stains on cloth while the anionic end is attracted by ion–dipole forces to water molecules. They are examples of anionic detergents.

fat soluble end   water soluble end

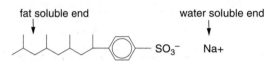

SO_3^- $Na+$

a detergent: more soluble than soap; calcium and magnesium salts are also soluble so no scum forms, but stable so forms foam in rivers

Dyes

Many derivatives of phenylbenzene are coloured and have been used as dyes. Some occur naturally, while others are synthesised artificially.

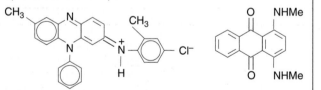

Aniline mauve: a natural dye   Alizarin: a synthetic dye

Drugs

Many well known drugs are based on aromatic compounds. Aspirin and paracetamol are both phenol derivatives.
The early anti-bacterial drugs were based on derivatives of phenylbenzene and were called sulphonamides.
Narcotics such as morphine, codeine, and heroin are all aromatic compounds.

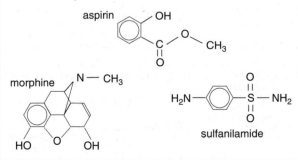

sulfanilamide

Aromatic hydrocarbons: the theory

STRUCTURE

There are three sigma bonds around each carbon atom, so they will be at 120° to each other. This makes benzene a planar (flat) molecule.

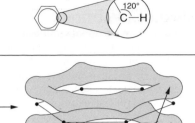

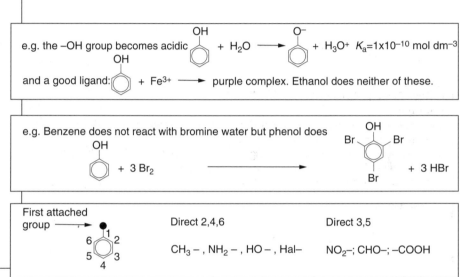

π overlap

each point • is one carbon atom bonded to a hydrogen atom

delocalised π orbitals above and below the σ skeleton

AROMATICITY

Aromatic substances have the following properties:

1. Their molecules are planar.

2. The reactivity of the atoms or groups of atoms joined to the aromatic ring are modified compared to their properties when joined to other carbon skeletons.

3. The reactivity of the aromatic ring is changed by the presence of an attached group — a substituent. These groups may make the ring more or less reactive than benzene itself.

4. Once one group has been joined to the ring, it directs further incoming groups to particular places on the ring.

e.g. the –OH group becomes acidic [structure] + H_2O ⟶ [structure] + H_3O^+ $K_a = 1 \times 10^{-10}$ mol dm⁻³

and a good ligand: [structure] + Fe^{3+} ⟶ purple complex. Ethanol does neither of these.

e.g. Benzene does not react with bromine water but phenol does

[phenol] + 3 Br_2 ⟶ [2,4,6-tribromophenol] + 3 HBr

First attached group ⟶ [ring numbered 1–6]

Direct 2,4,6

CH_3– , NH_2– , HO– , Hal–

Direct 3,5

NO_2–; CHO–; –COOH

REACTIVITY

Because the π electrons are delocalised, the electron density between any two atoms is diluted as the π electrons spread out over the ring. This makes aromatic systems more stable (see p. 52) and less reactive than expected. Aromatic systems only react with the strongest electrophiles or with the help of a catalyst.

[C=C structure] ← four electrons between atoms   [benzene structure] ← only three electrons between atoms

Nitration

When methylbenzene is nitrated, the attacking electrophile is the very reactive nitronium ion. The methyl group already on the ring directs the incoming nitro group to position 2, 4, or 6. If the temperature of the reaction is kept below 40 °C no more nitro groups will come in, because the first one deactivates the ring. This is just as well because further nitration makes trinitrotoluene, TNT!

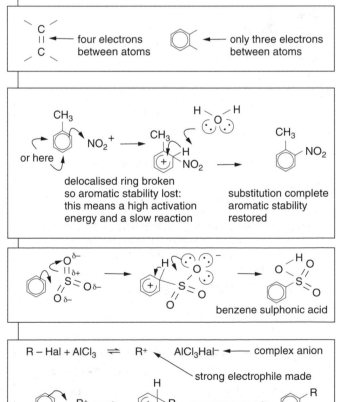

delocalised ring broken so aromatic stability lost: this means a high activation energy and a slow reaction

substitution complete aromatic stability restored

Sulphonation

The sulphur trioxide molecule is another strong electrophile which can react with benzene.

benzene sulphonic acid

Alkylation

The high charge density of the aluminium ion in a Friedel–Crafts catalyst gives it great polarising power, which helps to make the halogen compound into a strong enough electrophile to attack the benzene.

R – Hal + $AlCl_3$ ⇌ R^+ $AlCl_3Hal^-$ ← complex anion

strong electrophile made

[benzene] + R^+ ⟶ [intermediate] R ⟶ deprotonation by a base ⟶ [substituted benzene] R

Benzene derivatives: phenol and phenylamine: the facts

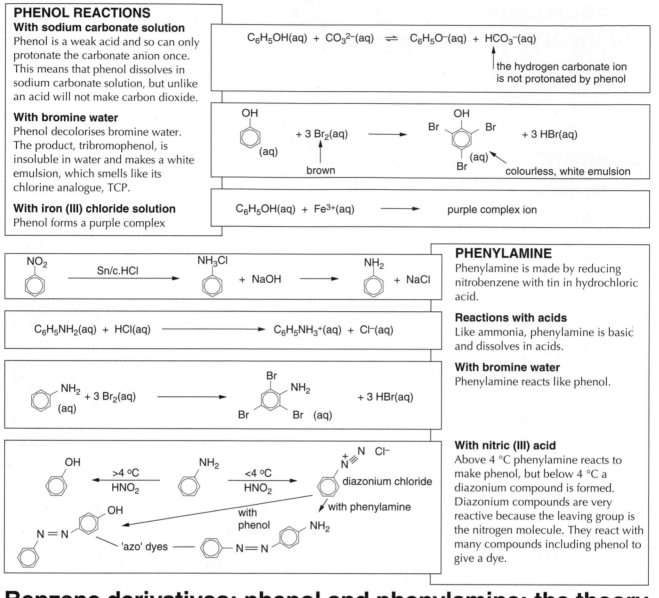

PHENOL REACTIONS

With sodium carbonate solution
Phenol is a weak acid and so can only protonate the carbonate anion once. This means that phenol dissolves in sodium carbonate solution, but unlike an acid will not make carbon dioxide.

$$C_6H_5OH(aq) + CO_3^{2-}(aq) \rightleftharpoons C_6H_5O^-(aq) + HCO_3^-(aq)$$

the hydrogen carbonate ion is not protonated by phenol

With bromine water
Phenol decolorises bromine water. The product, tribromophenol, is insoluble in water and makes a white emulsion, which smells like its chlorine analogue, TCP.

With iron (III) chloride solution
Phenol forms a purple complex

$$C_6H_5OH(aq) + Fe^{3+}(aq) \longrightarrow \text{purple complex ion}$$

$$C_6H_5NH_2(aq) + HCl(aq) \longrightarrow C_6H_5NH_3^+(aq) + Cl^-(aq)$$

PHENYLAMINE

Phenylamine is made by reducing nitrobenzene with tin in hydrochloric acid.

Reactions with acids
Like ammonia, phenylamine is basic and dissolves in acids.

With bromine water
Phenylamine reacts like phenol.

With nitric (III) acid
Above 4 °C phenylamine reacts to make phenol, but below 4 °C a diazonium compound is formed. Diazonium compounds are very reactive because the leaving group is the nitrogen molecule. They react with many compounds including phenol to give a dye.

Benzene derivatives: phenol and phenylamine: the theory

The reactivities of both phenol and phenylamine are good examples of aromaticity. Both the hydroxy and amino groups are modified by being joined to the benzene ring and the reactivity of the benzene ring is changed by the presence of the joined groups.

MODIFICATION OF THE HYDROXY GROUP

One of the lone pairs on the oxygen atom becomes conjugated with (joined to) the π system. This withdrawal of electrons from the oxygen makes the attached hydrogen more positive and so more readily attacked by lone pairs i.e. more acidic. Compare this aryl hydroxy compound with water, alkyl, and acyl ones:

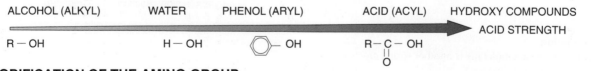

MODIFICATION OF THE AMINO GROUP

The lone pair becomes conjugated with the π electrons and so is much less available to acids. Phenylamine is therefore a much weaker base than ethylamine. Compare its base strength with that of ammonia, alkyl amines, and amides.

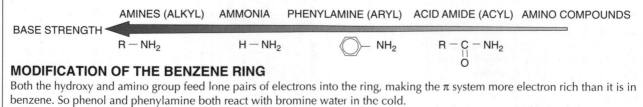

MODIFICATION OF THE BENZENE RING

Both the hydroxy and amino group feed lone pairs of electrons into the ring, making the π system more electron rich than it is in benzene. So phenol and phenylamine both react with bromine water in the cold.

Amino acids: the theory

There are many alpha-amino acids. They differ from each other in the side chain joined to the alpha carbon atom. This carbon atom has four different groups joined to it so it is optically active and there are two stereoisomers of each amino acid. In the proteins of all living things from bacteria to humans the same twenty amino acids are found linked together in protein chains. All these amino acids are the left-handed isomer.

Some of the common alpha-amino acids are

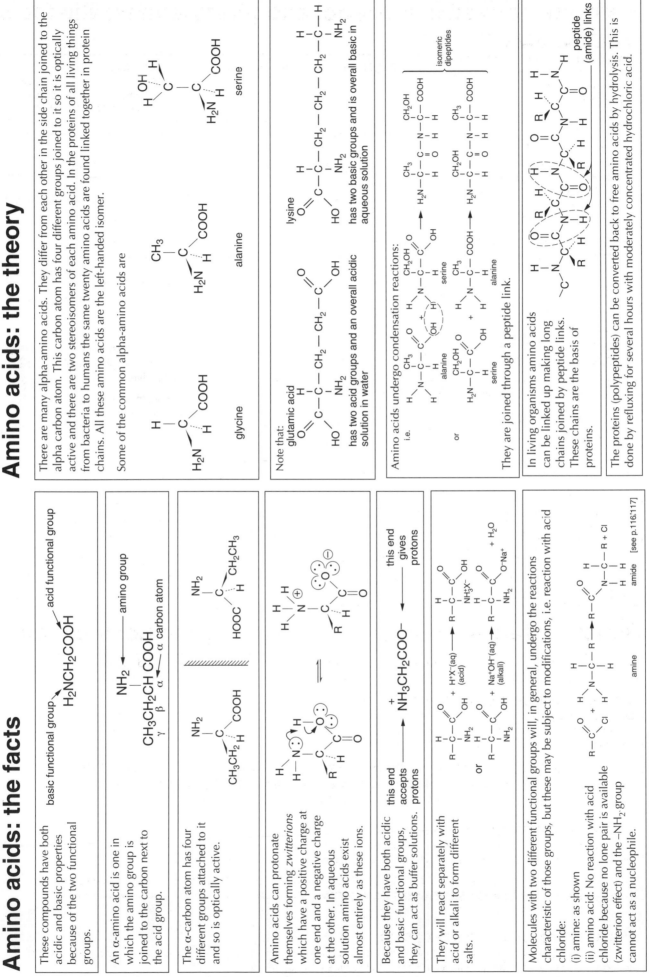

glycine

alanine

serine

Note that:

glutamic acid has two acid groups and an overall acidic solution in water

lysine has two basic groups and is overall basic in aqueous solution

Amino acids undergo condensation reactions:

i.e.

or

They are joined through a peptide link.

isomeric dipeptides

In living organisms amino acids can be linked up making long chains joined by peptide links. These chains are the basis of proteins.

peptide (amide) links

The proteins (polypeptides) can be converted back to free amino acids by hydrolysis. This is done by refluxing for several hours with moderately concentrated hydrochloric acid.

Amino acids: the facts

These compounds have both acidic and basic properties because of the two functional groups.

basic functional group → acid functional group

H_2NCH_2COOH

An α-amino acid is one in which the amino group is joined to the carbon next to the acid group.

NH_2 ← amino group

$CH_3CH_2CH\ COOH$
$\gamma\quad\beta\quad\alpha$ ← α carbon atom

The α-carbon atom has four different groups attached to it and so is optically active.

Amino acids can protonate themselves forming *zwitterions* which have a positive charge at one end and a negative charge at the other. In aqueous solution amino acids exist almost entirely as these ions.

Because they have both acidic and basic functional groups, they can act as buffer solutions.

this end accepts protons

$\overset{+}{N}H_3CH_2COO^-$

this end gives protons

They will react separately with acid or alkali to form different salts.

$+ H^+X^-$(acid)

$+ Na^+OH^-$(aq) (alkali)

$+ H_2O$

Molecules with two different functional groups will, in general, undergo the reactions characteristic of those groups, but these may be subject to modifications, i.e. reaction with acid chloride:

(i) amine: as shown

(ii) amino acid: No reaction with acid chloride because no lone pair is available (zwitterion effect) and the –NH_2 group cannot act as a nucleophile.

amine

$+ R + Cl$

amide   [see p.116/117]

Diagnostic tests for functional groups

Functional group	Reactive site	Distinguish by	What you have to do	What you see if the test is positive
Alkene	$C=C$	Electrophilic addition	Add bromine water to the substance and shake	The bromine water goes colourless
			Add acidified potassium manganate (VII) and shake	The manganate (VII) goes colourless
Carbonyl	$\overset{\delta+}{C}=\overset{\delta-}{O}$	Addition/elimination	Collect some 2,4-dinitrophenylhydrazine in a dry tube. Add a very small amount of the substance and shake	A yellow or orange precipitate forming slowly
Aldehyde	$H-C=O$	Reducing properties	Collect some silver nitrate solution. Add two or three drops of sodium hydroxide solution, then aqueous ammonia solution until the mixture is almost colourless. Now add a very small amount of the substance. Warm if necessary	A grey/black precipitate or silver mirror
Ethanoyl group or its precursor	$CH_3-C=O$	Addition reaction	Take some iodine in potassium iodide solution; add sodium hydroxide until it is pale yellow. Add a few drops of substance and shake	A yellow precipitate (iodoform) slowly forms
Primary/secondary alcohols	$-\overset{H}{\underset{|}{C}}-OH$	Reducing properties	Collect some potassium dichromate solution and add dilute sulphuric acid. Now add the substance and warm	On warming the colour changes from orange to brown then green
Acid	$\overset{O}{\underset{}{C}}-O-H$	Protonating properties	Add some sodium carbonate solution	Evolution of carbon dioxide
Acid chlorides	$\overset{O}{\underset{}{C}}-Cl$	Addition/elimination	Add the substance to water	Evolution of hydrogen chloride
Phenol	$O-H$ (phenol ring)	Protonating properties	Add the substance to sodium carbonate solution	Dissolves without the evolution of carbon dioxide
		Complexing properties	Add to water and shake; now add a drop or two of iron (III) chloride solution	A purple colour is seen
Phenol and phenylamine	:N or O (phenyl ring)	Electrophilic substitution	Add bromine water and shake	Bromine colour goes and a white ppt. is seen

Polymers I

DEFINITION

A **polymer** is a large molecule made by linking together many smaller ones. Smaller molecules can link up if:

- either they use electrons in a π bond to make new σ bonds to the next molecule, in which case the product is called an **addition polymer**

- or they eliminate atoms freeing some bonds to link to the next molecule; the atoms eliminated form a small well known molecule such as water or ammonia. The polymer formed in this case is called a **condensation polymer**.

The polymer may be a long chain or joined in two dimensions forming a sheet.

ADDITION POLYMERISATION

Here the small molecules, called **monomers**, have a double bond between two carbon atoms. Polymerisation is either:

1. triggered by a free radical: this is provided by adding a substance called an **initiator**. The sequence of the reaction is similar to the chain reaction between hydrocarbons and chlorine and is carried out at high pressure and temperature.

Initiation:
Peroxide initiator molecules break up making free radicals which react with monomer molecules starting a growing chain.

Propagation:
Here growing chains react with more monomers

Termination:
At this stage two growing chains meet and make an unreactive or 'dead' polymer.

or

2. carried out in the presence of a Ziegler catalyst, which is titanium chloride, under low pressure and temperature. The Ziegler catalyst acts as a molecular jig, holding the growing chain while another monomer molecule adds on.

EXAMPLES OF ADDITION POLYMERS

Monomer	Polymer
ethene	polythene
propene	polypropene
chloroethene (vinyl chloride)	polyvinylchloride (PVC)
phenylethene (styrene)	polystyrene
tetrafluoroethene	polytetrafluoroethene (Teflon non-stick)

Because addition polymers have no reactive sites, they are **non-biodegradable** and cause a disposal problem.

THE POLYMERISATION PROCESS

- **energy**: the polymerisation process is exothermic because it involves breaking a weaker π bond and making a stronger σ bond.

- **viscosity**: as polymerisation proceeds and the chains get longer, the van der Waals forces between them increase. This leads to an increase in viscosity of the reaction liquid.

Polymerisation plants are therefore designed to remove heat from the reaction vessel and to keep viscosity fairly constant either by adding an inert solvent or water in which the reacting particles become suspended.

breaking π bond $+ 264$ kJ mol^{-1}

making a new σ bond $- 348$ kJ mol^{-1}

Initiation:

$$C_6H_5C(O){-}O{-}O{-}CC_6H_5 \rightarrow 2C_6H_5COO\cdot \rightarrow 2C_6H_5\cdot + CO_2$$

$$C_6H_5\cdot + CH_2{=}CH_2 \rightarrow C_6H_5CH_2{-}CH_2\cdot \quad \text{a growing chain}$$

Propagation:

$$C_6H_5(CH_2)_x CH_2CH_2\cdot + CH_2{=}CH_2 \rightarrow C_6H_5(CH_2)_{x+2} CH_2{-}CH_2\cdot$$
a longer chain

Termination:

$$C_6H_5(CH_2)_x CH_2CH_2\cdot + \cdot CH_2CH_2(CH_2)_y C_6H_5 \rightarrow C_6H_5(CH_2)_x CH_2CH_2CH_2(CH_2)_y C_6H_5$$
unreative polymer chain

monomer held while the growing chain joins

new monomer and so on...

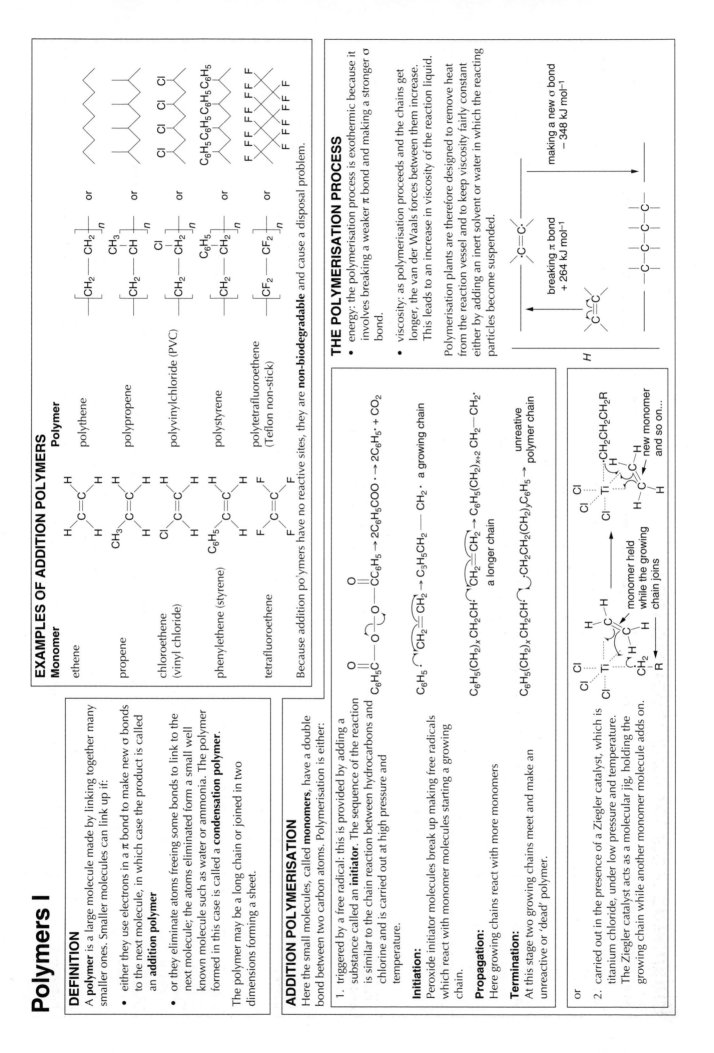

Polymers II

CONDENSATION POLYMERS

Nylon and polyester are examples of condensation polymers. They are given this name because when the small molecules link together, water or some other small, stable molecule is eliminated just as when acid reacts with alcohol.

A long chain can be formed because **bifunctional molecules** with a reactive site at each end can link up indefinitely.

$$CH_3COOH + HOCH_2CH_3 \rightleftharpoons$$

atoms eliminated to free bonds

$$CH_3COOCH_2CH_3 + H_2O$$

new ester link   small molecule also formed

Nylons are linked together by amide links and are sometimes called polyamides.

$$x\ HNH(CH_2)_6HNH + y\ ClC(CH_2)_4CCl$$

amide links

The nylon is numbered depending on how many carbon atoms there are between each amide link. The one above is formed from two small molecules each with six carbons in it and so is called nylon 6.6. Sometimes nylons are formed from a single kind of small molecule which has different functional groups at each end, for example nylon 6:

$$n\ HNH(CH_2)_5CCl$$

Polyesters are joined together by ester links. The acid used is benzene-1,4-dicarboxylic acid because polyesters made from the alkyl acids are too soluble. The commonest polyester is called terylene.

$$x\ HOOC-\bigcirc-COOH + y\ HOCH_2CH_2OH$$

$$HOOC-\bigcirc-COOCH_2CH_2OOC-\bigcirc-COOCH_2CH_2OOC-\bigcirc$$

CLASSIFYING PLASTICS

The plastics made from polymers are classified by the effect that heat has on them.

A **thermoplastic** can be heated until it softens, and then cooled to harden again, repeatedly. Addition polymers, nylons, and polyesters, are thermoplastics and are often melted in the process of manufacture into some useful material.

A **thermoset**, once heated, hardens as new covalent bonds form and will not soften again. Thermosets are formed by the crosslinking of the monomer molecules. The more the plastic is heated — cured — the greater the crosslinking and so the harder the plastic becomes.

Thermosets are used for light fittings, saucepan handles, and similar objects which have to survive repeated abrasion during usage. A common thermoset is made from phenol and methanal. The hydroxy group activates the hydrogens on the 2, 4, and 6 positions on the benzene ring so allowing bonds to form in more than one dimension. A crosslinked polymer forms.

activated hydrogen on phenol

$$+ H_2O \quad \text{and so on making a chain}$$

Heating (called curing) makes cross-links (covalent bonds) between the chains:

a 'cured' thermoset plastic: very hard

Protein synthesis – nucleic acids and sugars

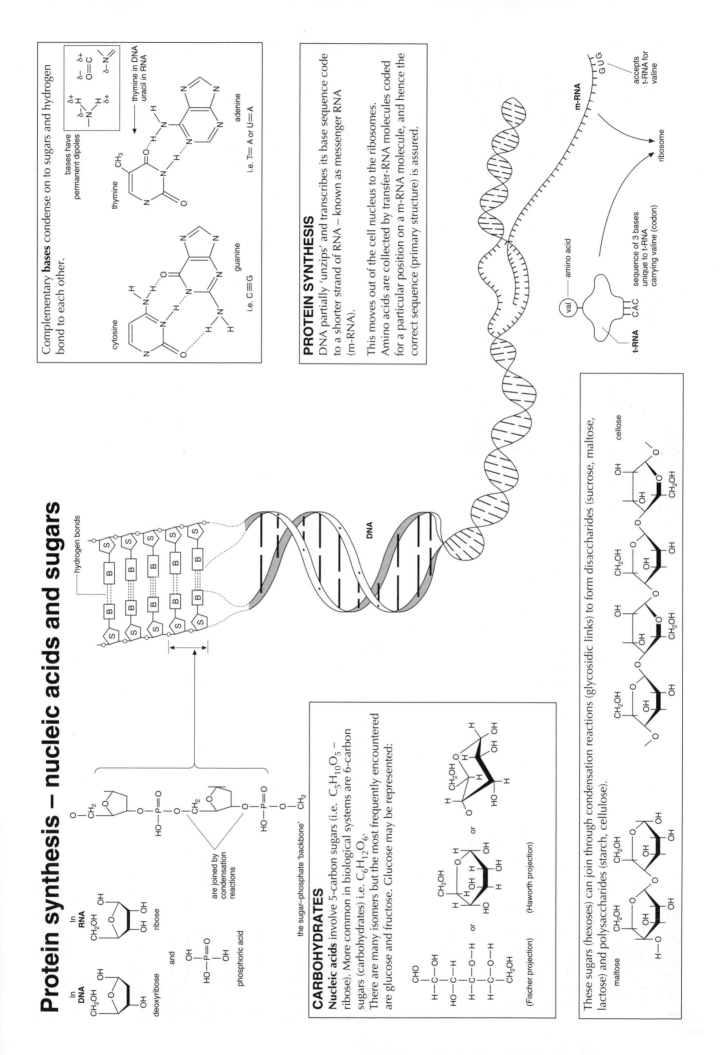

Complementary **bases** condense on to sugars and hydrogen bond to each other.

bases have permanent dipoles

thymine in DNA uracil in RNA

thymine

adenine

i.e. T= A or U=A

cytosine

guanine

i.e. C≡G

PROTEIN SYNTHESIS

DNA partially 'unzips' and transcribes its base sequence code to a shorter strand of RNA – known as messenger RNA (m-RNA).

This moves out of the cell nucleus to the ribosomes. Amino acids are collected by transfer-RNA molecules coded for a particular position on a m-RNA molecule, and hence the correct sequence (primary structure) is assured.

m-RNA

GUG

accepts t-RNA for valine

ribosome

amino acid

val

sequence of 3 bases unique to t-RNA carrying valine (codon)

CAC

t-RNA

DNA

hydrogen bonds

CARBOHYDRATES

Nucleic acids involve 5-carbon sugars (i.e. $C_5H_{10}O_5$ – ribose). More common in biological systems are 6-carbon sugars (carbohydrates) i.e. $C_6H_{12}O_6$.
There are many isomers but the most frequently encountered are glucose and fructose. Glucose may be represented:

(Fischer projection)

CHO
H—C—OH
HO—C—H
H—C—O—H
H—C—O—H
CH₂OH

or

(Haworth projection)

In DNA
deoxyribose

CH₂OH OH
OH

and

In RNA
ribose

CH₂OH OH
OH OH

phosphoric acid

OH
HO—P—OH
O

are joined by condensation reactions

the sugar–phosphate 'backbone'

These sugars (hexoses) can join through condensation reactions (glycosidic links) to form disaccharides (sucrose, maltose, lactose) and polysaccharides (starch, cellulose).

maltose

cellose

Protein structures – a study in intermolecular bonding

See 'Amino acids'.

Proteins consist of 50 or more (often hundreds) **amino acids**, held by peptide links. The exact sequence is called the **primary structure.**

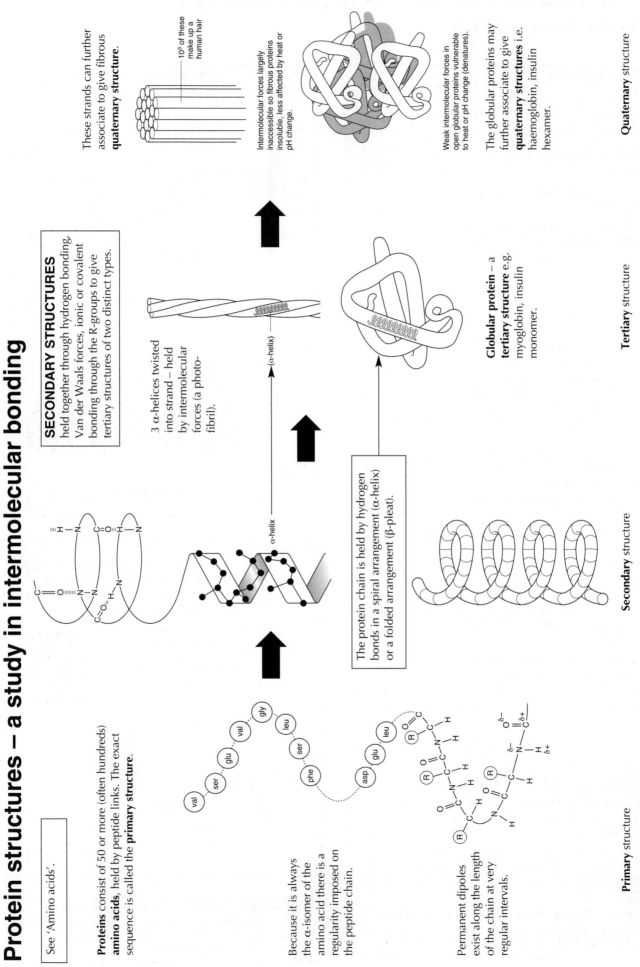

Because it is always the α-isomer of the amino acid there is a regularity imposed on the peptide chain.

Permanent dipoles exist along the length of the chain at very regular intervals.

SECONDARY STRUCTURES
held together through hydrogen bonding, Van der Waals forces, ionic or covalent bonding through the R-groups to give tertiary structures of two distinct types.

3 α-helices twisted into strand – held by intermolecular forces (a photo-fibril).

(α-helix)

α-helix

The protein chain is held by hydrogen bonds in a spiral arrangement (α-helix) or a folded arrangement (β-pleat).

These strands can further associate to give fibrous **quaternary structure.**

10³ of these make up a human hair

Intermolecular forces largely inaccessible so fibrous proteins insoluble, less affected by heat or pH change.

Weak intermolecular forces in open globular proteins vulnerable to heat or pH change (denatures).

The globular proteins may further associate to give **quaternary structures** i.e. haemoglobin, insulin hexamer.

Globular protein – a **tertiary structure** e.g. myoglobin, insulin monomer.

Primary structure

Secondary structure

Tertiary structure

Quaternary structure

Raw materials for the chemical industry I

FORMATION OF ORE BODIES

Minerals and ores

There is not an even distribution of the elements throughout the Earth. Some elements will tend to associate with the lighter silicate crust, others with the denser mantle, yet others within the crust. Geological events have brought about changes in this distribution. Naturally occurring compounds are called **minerals**. Rocks are mixtures of minerals.

Any mineral mass than can be exploited profitably is referred to as an ore.

To become of commercial interest, its mineral must undergo some form of concentration process, through reorganisable geological events.

Hydrothermal deposits

(a) The magma has a high water content, effectively superheated steam. Many metals will exist as hydrated ions in solution. As the magma cools silicate minerals gradually crystallise out, leaving a very mobile fluid containing metal ions together with sulphide and oxide ions.

This hot, aqueous 'soup' will be intruded into cracks and weaknesses in the surrounding rock (as a result of increasing pressure). There they will cool, and, depending upon solubility the metal sulphides (or oxides) will crystallise out.

The resultant mineral concentrations are therefore referred to as **hydrothermal deposits** (e.g. ZnS, PbS, $CuFeS_2$, Ag_2S, SnO_2).

(b) As plates move apart the rocks are distributed. Water can penetrate and dissolve up metal ions, return to the colder ocean floor and deposit metal sulphides. These too are referred to as hydrothermal deposits (ZnS, PbS, $CuFeS_2$, etc.).

NB During slow cooling of certain types of magma, heavy, simple minerals may form and accumulate at the bottom. This is magmatic segregation and results in Fe_3O_4, $FeCr_2O_4$.

Evaporite deposits

Solible metal ions (Na^+, K^+, Ca^{2+}, Mg^{2+}) are washed into the sea, together with halide ions, sulphate ions, and carbonate ions.

Shallow seas in hot climates evaporate quickly, concentrating the solution of ions, until the point of saturation of certain compounds is reached and they start to crystallise out.

Over geological time this has given rise to thick deposits of rock salt or halite ($NaCl$), gypsom $CaSO_4.2H_2O$, anhydrite $CaSO_4$, and mixed halides. These are **evaporite** deposits.

The **theory of plate tectonics** describes the Earth's crust as a number of interlocking, mobile plates moving towards each other or away from each other at a rate of a few centimetres a year.

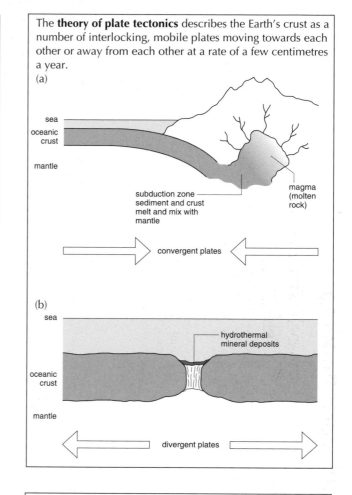

Placer deposits

Geological upheavals may bring these mineralised veins close to the surface. Here they are subject to erosion by running water. The eroded material is carried downstream in solution or suspended in fast-moving water.

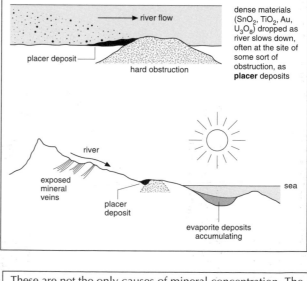

These are not the only causes of mineral concentration. The mobility of aqueous metal ions during various sedimentary processes is thought to be the cause of some of the most valuable iron and copper ore bodies.

Raw materials for the chemical industry II

EXPLOITATION

Apart from iron ore, most of the UK's metal requirements are imported. Small deposits of copper, lead, zinc, and tin are now worked out.

For economic reasons primary processing is carried out to the mine. The import is mineral concentrate.

The sort of mine (open-pit or shaft) depends on factors such as size, depth and grade of ore, together with environmental considerations.

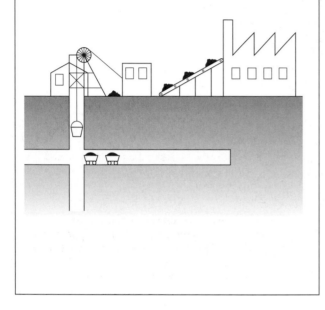

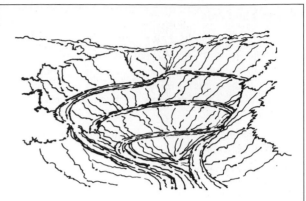

Ore processing
Whether the mine is open-pit or shaft, the ore body will probably be blasted to a rubble with explosive. The content of the required metal compound is probably quite low (often about 0.5% for copper ores), and the mineral particles are small.

The first stage, common to many ores, is to 'liberate' the mineral particles. The rock is crushed and ground in huge rotating ball mills to about sand-grain size.

What happens next will depend upon a number of factors – the chemical nature, concentration, and local environment amongst them.

The common feature is that ores are concentrated, smelted (treated chemically), to obtain in element form, then refined.

- See 'Metal extraction' for details of extraction of Al, Cr, Zn, and Fe from concentrates.

Case study – copper
Depending on the nature of the starting ore material there are two entirely different processing routes:

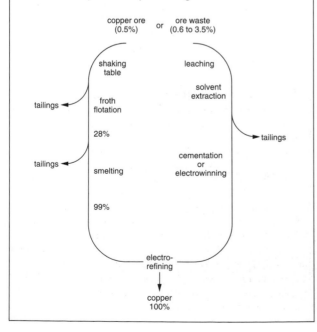

The minerals industry has its own terminology:

Leaching – getting metal ions into solution (using acids, alkalis, or complexing agents) i.e.
$$Cu^{2+} \rightarrow Cu^{2+}(aq) \text{ or } Cu(NH_3)_4(H_2O)_2^{2+}(aq)$$

Solvent extraction – use of organic solvent to selectively remove certain ions from impure aqueous solution i.e.
$$Cu^{2+}(aq) + \text{ligand (organic)} \rightarrow Cu(\text{ligand})^{2+}(\text{organic})$$

Cementation – displacement by another metal from aqueous solution i.e.
$$Cu^{2+}(aq) + Zn \rightarrow Cu(s) + Zn^{2+}(aq)$$

Electrowinning – using electrolysis of aqueous solution to remove metal i.e. $Cu^{2+}(aq) + 2e^- \rightarrow Cu(s)$

Electrorefining – purification by electrolysis (impure copper as anode)
$$Cu(s) \rightarrow Cu^{2+}(aq) + 2e^-; \ Cu^{2+}(aq) + 2e^- \rightarrow Cu(s)$$

Smelting – redox reactions in molten mixtures i.e.
$$4CuFeS_2 + 10\tfrac{1}{2}O_2 \rightarrow 4Cu + 2FeO + Fe_2O_3 + 8SO_2$$

Froth flotation – copper minerals selectively removed by adhering to stream of air bubbles, having been conditioned with 'collector' chemical.

Shaking table – device for continuous separation by density difference.

Tailings – waste product (sand size or smaller) carried by water to tailings ponds to settle out.

The chemical industry

UK CHEMICAL INDUSTRY

- Accounts for 2.5% of Gross National Product
- Accounts for about 11% total manufacturing gross value
- Sales worth £32 billion (late 1990s)
- Overseas sales yielded trade surplus about £4 billion
- Royalties and 'know-how' payments exceed £1.25 billion
- Largest single sector is pharmaceuticals, followed by plastics, then soaps/detergents.

FEEDSTOCK/RAW MATERIALS

- Reactants for main chemical process called **feedstock**
- **Feedstock** produced from **raw materials** (purification or extensive chemical treatment) (Methane is raw material to produce hydrogen gas as feedstock for Haber process (see p. 92)
- Size or state may need adjustment for ease of transportation
- Energy a raw material, as is air and water.

LOCATION

The UK chemical industry is concentrated in certain parts of the country for good reasons (some by now historical). These might include:

- proximity to feedstock/raw material supply
- proximity to large market
- deep sea access/suitable port facilities (for import and for export)
- good communications (road and rail) infrastructure
- level site with room for expansion
- constant water supply
- access to work force
- qualification for subsidy.

COSTS

Like any other manufacturing industry, the chemical industry sets out to generate profit.

- Research and development costs that precede manufacture can be immense, particularly in the world of pharmaceuticals.

Production costs are seen as two components:

- **Fixed (indirect) costs** are those incurred irrespective of the amount of product produced. These include plant depreciation (**capital costs**); land purchase; rates and services; salaries.
- **Variable costs** depend upon the extent of production – i.e. raw materials; distribution; wages; effluent treatment.

Selling price will depend upon production costs and needs to allow for preproduction and profit.

PROCESSING CONDITIONS

To be profitable the process should give the maximum yield in the shortest possible time, with as few stages as possible. In effect this means a compromise between kinetic considerations (pp. 54–55) and those of equilibrium (pp. 58–63)

- Temperature control also has energy cost implications
- Pressure increases could require costly pipes and pumps
- Catalysts may be expensive and have low life expectancy
- Co-products can increase profitability, but by-products can increase disposal costs
- Safety considerations are governed by strict legislation.

FROM LABORATORY TO PLANT

- Research and development is a small-scale, laboratory-based activity.
- A selected process is tested on a scale down version of the proposed plant. This is the **pilot plant**.
- Teething problems overcome, the process can then be transferred to the final manufacturing plant.

BATCH OR CONTINUOUS PROCESS

- **Batch processing** means raw materials mixed together in a vessel and reaction allowed to proceed, with monitoring.
  On completion products removed and separated. Vessel, after cleaning, can be used again, or slightly modified for a different product. Ideal for low-demand products, or if the reaction takes time. Plant costs are low but labour costs higher, and contamination risks are high.
- In a **continuous process** raw materials are mixed and react as they flow through the plant (large tanks or long tubes), and emerge at the end for processing. Well suited to high demand products, especially if the process is susceptible to automatic control. High plant costs, but low labour costs. The plant is less flexible and feedstock variability can be a problem.

ENVIRONMENTAL ISSUES

The chemical industry is sensitive to such concerns and is subject to regulation and legislation. In general terms this is with regard to:

- emissions (strict monitoring together with technological improvement minimise these)
- effluent treatment – particularly extracted water
- waste disposal
- transport of chemicals

Any breaches of legislation should be through accident alone.

Environmental chemistry

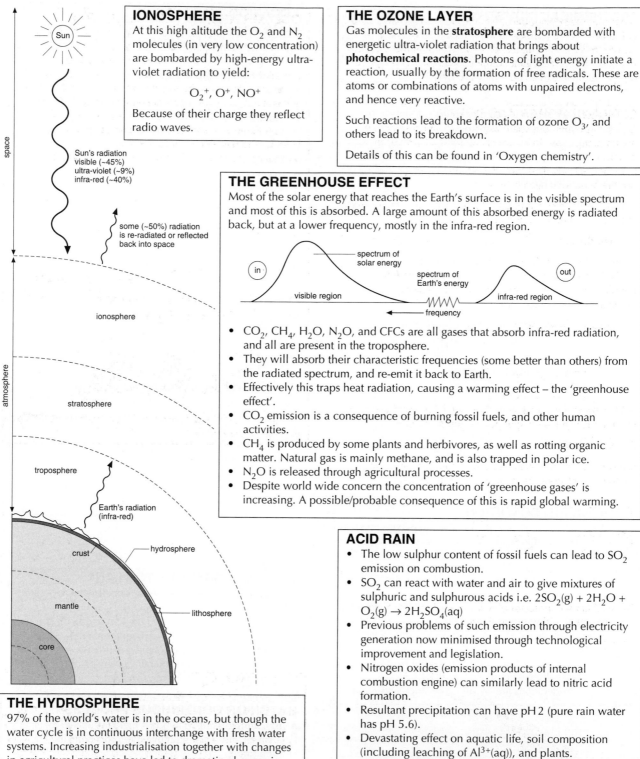

IONOSPHERE

At this high altitude the O_2 and N_2 molecules (in very low concentration) are bombarded by high-energy ultra-violet radiation to yield:

$$O_2^+, O^+, NO^+$$

Because of their charge they reflect radio waves.

THE OZONE LAYER

Gas molecules in the **stratosphere** are bombarded with energetic ultra-violet radiation that brings about **photochemical reactions**. Photons of light energy initiate a reaction, usually by the formation of free radicals. These are atoms or combinations of atoms with unpaired electrons, and hence very reactive.

Such reactions lead to the formation of ozone O_3, and others lead to its breakdown.

Details of this can be found in 'Oxygen chemistry'.

THE GREENHOUSE EFFECT

Most of the solar energy that reaches the Earth's surface is in the visible spectrum and most of this is absorbed. A large amount of this absorbed energy is radiated back, but at a lower frequency, mostly in the infra-red region.

- CO_2, CH_4, H_2O, N_2O, and CFCs are all gases that absorb infra-red radiation, and all are present in the troposphere.
- They will absorb their characteristic frequencies (some better than others) from the radiated spectrum, and re-emit it back to Earth.
- Effectively this traps heat radiation, causing a warming effect – the 'greenhouse effect'.
- CO_2 emission is a consequence of burning fossil fuels, and other human activities.
- CH_4 is produced by some plants and herbivores, as well as rotting organic matter. Natural gas is mainly methane, and is also trapped in polar ice.
- N_2O is released through agricultural processes.
- Despite world wide concern the concentration of 'greenhouse gases' is increasing. A possible/probable consequence of this is rapid global warming.

THE HYDROSPHERE

97% of the world's water is in the oceans, but though the water cycle is in continuous interchange with fresh water systems. Increasing industrialisation together with changes in agricultural practices have led to dramatic changes in fresh water quality. Monitoring and treatment procedures for domestic supplies are subject to stringent legislation.

THE LITHOSPHERE

This is the outer part of the mantle and the crust and is in the main rock made up of silicate minerals. Soil results from erosion processes, and soil chemistry determines agricultural practice. Demand for food has meant increasing use of artificial fertilisers (and a consequent excess of nirate in water supplies). The more widespread use of pesticides and herbicides is giving rise to excessive organic residues with potentially harmful effect.

ACID RAIN

- The low sulphur content of fossil fuels can lead to SO_2 emission on combustion.
- SO_2 can react with water and air to give mixtures of sulphuric and sulphurous acids i.e. $2SO_2(g) + 2H_2O + O_2(g) \rightarrow 2H_2SO_4(aq)$
- Previous problems of such emission through electricity generation now minimised through technological improvement and legislation.
- Nitrogen oxides (emission products of internal combustion engine) can similarly lead to nitric acid formation.
- Resultant precipitation can have pH 2 (pure rain water has pH 5.6).
- Devastating effect on aquatic life, soil composition (including leaching of $Al^{3+}(aq)$), and plants.
- Metals and limestone undergo corrosion – deleterious effects on buildings, etc.

PHOTOCHEMICAL SMOG

- Traffic emissions combined with sunlight and certain geographic constraints can lead to photochemical smog, affecting cities such as Los Angeles and Athens.
- Nitrogen oxides and unburnt hydrocarbon react with oxygen in a genes of free-radical reactions leading to the formation of ozone, aldehydes, and the severely irritating peroxyethanoyl nitrate (PAN).

Colour chemistry

Most chemical compounds, whether ionic or covalent, are colourless. Colour comes about when an incomplete visible spectrum is radiated by a substance. In the main this is associated with d-block compounds and certain types of organic molecule. If soluble they are **dyes**.

CHROMOPHORES AND DYES

Groupings within organic compounds that cause colour are chromophores. They are characterised by delocalised electron systems together with functional groups with lone pairs of electrons. This includes alternating single and double bond (conjugated) systems.

For such coloured compounds to be useful dyes they need to be able to bond to another material.

(a) **Azo dyes** have 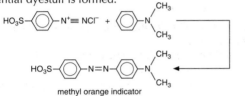 groups within them.

So when colourless benzene diazonium chloride (or similar) (see 'Benzene derivatives') combines with (i.e. couples with) another aromatic compound a coloured, potential dyestuff is formed:

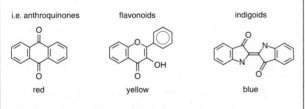

methyl orange indicator

NB Changing the chemical environment of the chromophore by introducing different groups changes the colour.

(b) **Natural dyes**

Many of the natural dyes that give colour to flowers, fruits, etc. have chromophoric groups involving oxygen:

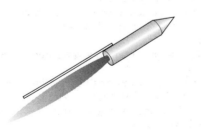

i.e. anthroquinones	flavonoids	indigoids
red	yellow	blue

The well known indicator phenolphthalein in its red form has similarities with anthroquinones.

FIREWORKS

Ignition of gunpowder mixed with certain s-block compounds gives conditions for excitation of electrons to higher energy levels. This energy is emitted as the characteristic colour as electrons return to lower levels (see 'ultraviolet and visible spectra') e.g. crimson colour from strontium salts. The incandescent sparkle from white hot metal filings (Fe, Al) adds to the effect.

COLOUR AND D-BLOCK ELEMENTS

The colour will depend on the metal, its oxidation state, and the nature of ligands bonded on to it. The colour comes about through absorption of certain visible frequencies by d-orbital electrons, as a consequence of complex formation. When uncomplexed the five d orbitals have similar energy levels, but ligand approach causes them to split. Electron transfer between the split d levels gives rise to colour. The exact frequency absorbed will depend on the nature of the ligand. Zinc has all d orbitals filled up, so electron promotion is not possible, and zinc compounds are colourless.

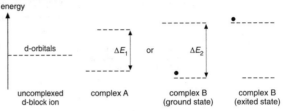

This is well illustrated by the absorption spectrum for $Ti(H_2O)_6^{3+}$ which is a purple colour.

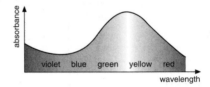

Blue, green, yellow are absorbed by promotion of the one d electron in the complex.

COLORIMETRY

Colorimetry is an analytical technique whereby the intensity of colour is used to measure concentration. The more a sample absorbs incident light, the less is transmitted. For a standard thickness of sample the intensity of transmitted light is related to concentration of coloured species. The intensity is measured on a meter connected to a photocell. For greater sensitivity a filter of complementary colour is used.

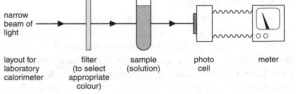

layout for laboratory calorimeter — filter (to select appropriate colour) — sample (solution) — photo cell — meter

For practical purposes the instrument needs to be calibrated using known concentrations.

Modifying properties – using metals

See 'Metallic bonding'; 'Metal extraction'.

METAL STRUCTURES

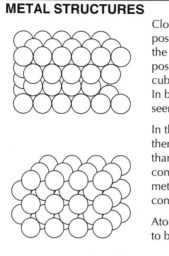

Close packing – maximum possible contact of spheres of the same size. There are two possible arrangements, one cubic, the other hexagonal. In both cases metal atoms are seen to be aligned.

In this cubic arrangement there is regularity, but less than the maximum possible contact between atoms. The metal is less dense as a consequence.

Atoms are nevertheless seen to be 'lined up'.

Crystals will grow as molten metal cools, and eventually form a mass of poorly shaped, interlocking crystals, usually referred to as **grains**. The points of contact of grains are 'grain boundaries'.

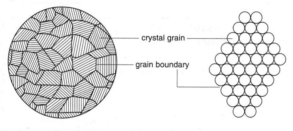

crystal grain

grain boundary

DISLOCATIONS

If there is a fault in the crystallisation such that some atoms are 'missing' then neighbouring planes of atoms are slightly displaced. In practice these faults (dislocations) are very common.

Under stress these dislocations can move through a structure, more easily than slip. It is the mobility of dislocations that gives a metal ductility.

Dislocations cannot cross grain boundaries, so crystal size (dependent on rate of cooling) makes a difference. Dislocations tend to interfere with each other and on meeting hinder further movement.

Activities such as:

work-hardening
annealing
hot working
recrystallisation
tempering

all maximise or minimise dislocation/slip, and thus affect a metal's properties.

PROPERTIES AND STRUCTURE

The mechanical properties of metals are what have made them so important over a long period of time. Strength, ductility and the ability to deform elastically are dependent on the metal itself, and the processing after extraction. Although the metallic bond (p. 31) is strong, it is not directional. This means that layers of atoms can slide over each other – thus metals can be hammered into new shapes.

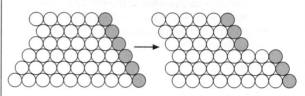

But this sliding (slip) is terminated at grain boundaries, so control of grain size will control mechanical properties.

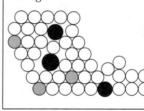

grain boundary

grain boundary

ALLOYS

Addition of atoms of different elements can also modify properties. If the atoms are larger then the sliding of layers is prevented, leading to a harder, less malleable metal. If the atoms are of similar size then there may be little change.

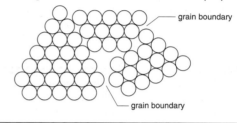

Dark circles represent larger atoms.

Hatched circles represent atoms of similar size, substituting in the crystal structure.

ALLOYS – SOME EXAMPLES

Steel is Fe with carbon plus other elements

Fe + 13% Mn give toughness

Fe + 20% Cr + 10% Ni gives stainless steel

Fe + 18% W + 5% Cr gives 'high-speed' steel

Cu + Zn is brass

Cu + Sn is bronze

Cu + 25% Ni is 'coinage silver'

Pb + Sn is pewter

Modifying properties – using plastics

See 'Polymers'.

Addition polymerisation generally yields a tangle of long, possibly branched molecules. Intermolecular bonding is possible at the points of contact. This means that molecules can move over each other when subject to heat or force, and any object fashioned from them can change shape. These are referred to as thermoplastics. Many instances of **condensation polymerisation** have scope for cross linking through covalent bonds. These will not readily yield with heat and force, the object will maintain its shape, and the term 'thermosetting plastic' is used.

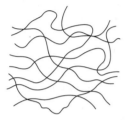 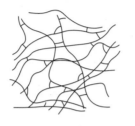

Thermoplastic
Intermolecular bonding may be van der Waals (instantaneous dipole – induced dipole) or permanent dipole – permanent dipole.

Thermosetting plastic
In this example the cross linking covalent bonds are randomly distributed.

Plastics are designed to meet exacting specifications, and several of the following approaches will in practice be employed in meeting customer requirements:
- copolymerisation
- chain length
- regularity within a chain
- degree of crosslinking
- use of plasticisers.

Designer plastics
Careful consideration of the chemical properties of monomers and intermolecular bonding potential has made it possible to produce plastics that are:

- electrical conductors – i.e. poly(pyrrole)
- photoconductors   – i.e. poly(vinyl carbazole)
- piezoelectric   – i.e. poly(1,1-difluorothene)
- soluble in water   – i.e. poly(ethanol)
- heat resistant   – i.e. poly(ether-ether-ketone)
- brodegradable   – i.e. poly(3-hydroxybutanoic acid)
- very strong   – i.e. Kevlar.

The greater the amount of intermolecular bonding, the stronger the material will be, and the less flexible. The degree of branching (which will limit contact) and the degree of alignment (referred to as **crystallinity**) will thus monitor properties.

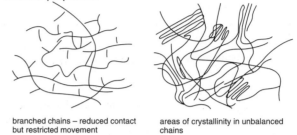

branched chains – reduced contact but restricted movement

areas of crystallinity in unbalanced chains

Crystallinity can be increased by aligning crystalline areas through 'cold-drawing' – stretching at room temperature.

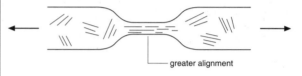

greater alignment

This also increases the density, giving rise to high- and low-density polymers.

Poly(ethene) is a good example of this.

The alternative approach to achieving optimum properties is to use a **plasticiser**. In this instance, chains that might normally have strong attractions (PVC for instance) can be kept apart by the introduction of small molecules, giving greater flexibility.

aligned molecules kept apart by plasticiser molecules e.g. flexible PVC

closely packed aligned molecules i.e. unplacticised PVC (uPVC)

Composites maximise the properties of two separate materials:

i.e. carbon fibres
polythene fibres
Kevlar fibres
glass fibres } in a matrix of epoxy resin or polyester

Revision questions

Questions are grouped as in section headings.

CHEMICAL CALCULATIONS

1   When 1.85 g potassium nitrate KNO_3 is heated it gives 1.56 g of pale yellow solid and 217.5 cm^3 of a colourless gas.

  (i)   How many moles of potassium nitrate are being heated?

  (ii) How many moles of gas are given off? (One mole of gas occupies 24 dm^3 at lab. conditions.)

2   In carrying out the following hydrolysis reaction:
$$C_6H_5COOCH_3 + H_2O \rightarrow C_6H_5COOH + CH_3OH$$
a student obtained 1.50 g benzoic acid, having started with 2.7 g methyl benzoate.

  (i)   How many moles of ester were used?

  (ii) What is the percentage yield?

3   1.40 g iron were made up to 100 cm^3 $FeSO_4$ solution in acid. 10.0 cm^3 of this solution required 24.2 cm^3 of 0.200 molar potassium manganate(VII) solution on titration. The reaction is:
$$5Fe^{2+}(aq) + MnO_4(aq) + 8H^+(aq)$$
$$\rightarrow 5Fe^{3+}(aq) + Mn^{2+}(aq) + 4H_2O(l)$$

  (i)   calculate the mass of iron in the steel sample

  (ii) calculate the percentage by mass of iron in steel.

4   Use the ideal gas equation to calculate the relative molecular mass of a gas, a 0.100 g sample of which occupies 52.9 cm^3 at a temperature of 377 K and a pressure of 1 atm. (1.01×10^5 N m^{-2}). ($R = 8.3$ J K^{-1} mol^{-1}.)

ATOMIC STRUCTURE AND SPECTRA

5   Account for the characteristic yellow colour emitted by sodium compounds when subjected to the flame test.

6   Use the following data to calculate the minimum frequency of radiation needed to break a single C—Br bond:
$$E(C—Br) = +290 \text{ kJ mol}^{-1}$$
Planck's constant (h) = 6.63×10^{-34} J Hz^{-1}
Avogadro constant (L) = 6.02×10^{23}

7   Give the full electronic structure for:
an atom of iron;
the Fe^{2+} ion.

8   What is the relationship between ^{12}C, and ^{14}C?

9   Complete the following equations:
$$^{226}_{88}Ra \rightarrow \quad ^{222}_{86}Rn \quad +$$
$$^{238}_{94}Pu \rightarrow \quad \quad + \quad ^{4}_{2}He$$
$$^{90}_{38}Sr \rightarrow \quad ^{90}_{39}Y \quad +$$
$$\rightarrow \quad ^{131}_{54}Xe \quad + \quad ^{0}_{-1}e$$

10 A student attempted to hydrolyse chlorobutane with water. After fifteen minutes of refluxing he withdrew a sample for infra-red spectroscopy analysis, which resulted in the trace shown above right. Using a data book, decide whether or not he had been successful. Give reasons for your answer.

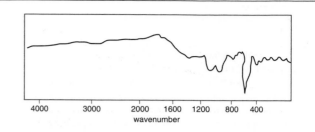

4000   3000   2000   1600   1200   800   400
wavenumber

11 Using a data book identify the three different proton environments distinguished on the nuclear magnetic resonance spectrogram below. Give the number of atoms in each environment and hence suggest a structure for the compound.

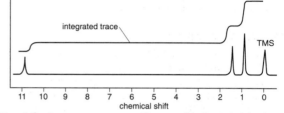

integrated trace

TMS

11   10   9   8   7   6   5   4   3   2   1   0
chemical shift

12 Use the mass spectrum illustrated below to calculate the relative atomic mass (A_r) for naturally occurring lead:

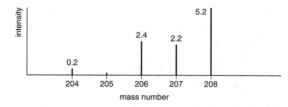

intensity

5.2

2.4   2.2

0.2

204   205   206   207   208
mass number

STRUCTURE AND BONDING

13 Draw 'dot-and-cross' diagrams for:
Strontium chloride, methanol,
the compound formed between boron trichloride and ammonia.
Comment on the type of bond in each instance.

14 Use diagrams to illustrate the shape of:
CH_4, SF_6, H_2O, PF_5, C_2H_4, H_3O^+
In each case give the bond angles.

15 Suggest a reason why aluminium oxide has an ionic structure, but aluminium chloride is covalent.

16 Why is the boiling temperature of chloromethane (CH_3Cl) so much higher than that of methane (CH_4)?

17 Why is poly(chloroethene) a more rigid plastic than poly(ethene)?

18 Give structural/displayed formulae and names for the isomers of pentane C_5H_{12}.

19 Give structural/displayed formulae and names for the isomers of but-2-ene.

20 The amino acid glycine $CH_2(NH_2)COOH$ has only one form, but alanine $CH_3CH(NH_2)COOH$ has two forms. Explain this using diagrams.

21 Describe, with the aid of a labelled diagram, a simple model of metallic bonding.

22 Successive ionisation energies of fluorine are shown on the plot below. By reference to fluorine's electron structure, explain the shape of the plot.

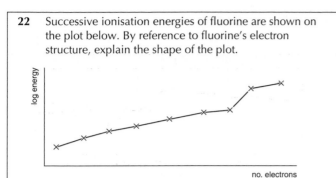

PHASE EQUILIBRIA

(NB Much of the content of this section is only applicable to certain options within certain syllabuses).

23 Write an equation to represent the condensation of water vapour. Using a kinetic model, give a brief qualitative description of condensation of a vapour, caused by a decrease in temperature.

THERMODYNAMICS

24 Given the following data, calculate the lattice enthalpy of rubidium bromide:

$\Delta H^{\ominus}_{\text{atomisation}}$(rubidium) $\quad = +80.9\,\text{kJ mol}^{-1}$
1st ionisation energy (rubidium) $= +403\,\text{kJ mol}^{-1}$
$\Delta H^{\ominus}_{\text{formation}}$(RbBr) $\quad\quad = -394.6\,\text{kJ mol}^{-1}$
$\Delta H^{\ominus}_{\text{atomisation}}$(bromine) $\quad = +111.9\,\text{kJ mol}^{-1}$
Electron affinity (bromine) $\quad = -324.6\,\text{kJ mol}^{-1}$

25 Given the following data, construct an enthalpy cycle diagram and use it to calculate a value for the standard enthalpy of formation of cyclohexane.

$\Delta H^{\ominus}_{\text{form}}(H_2O) \quad = -286\,\text{kJ mol}^{-1}$
$\Delta H^{\ominus}_{\text{form}}(CO_2) \quad = -394\,\text{kJ mol}^{-1}$
$\Delta H^{\ominus}_{\text{comb}}(C_6H_{12}) = -3924\,\text{kJ mol}^{-1}$

26 Given the following data, calculate a value for the bond energy $E(N—H)$.

$\Delta H^{\ominus}_{f}(NH_3) \quad\quad = -46.2\,\text{kJ mol}^{-1}$
$\Delta H^{\ominus}_{\text{atomisation}}(N_2) \quad = +473\,\text{kJ mol}^{-1}$
$\Delta H^{\ominus}_{\text{atomisation}}(H_2) \quad = +218\,\text{kJ mol}^{-1}$

27 What is meant by the term 'standard enthalpy of combustion' of a substance?

In an experiment to find a value for the enthalpy change of combustion of methanol, 1.92 g of the alcohol were burnt, raising the temperature of 200.00 g of water by 32.2 °C. Use the information to obtain a value.

28 Construct an enthalpy diagram to show the relationship between lattice energy, hydration energy, and enthalpy of solution for magnesium sulphate. Given the data:

lattice enthalpy $(MgSO_4)$ $\quad = -2833\,\text{kJ mol}^{-1}$
hydration enthalpy (Mg^{2+}) $\quad = -1891\,\text{kJ mol}^{-1}$
hydration enthalpy (SO_4^{2-}) $\quad = -1004\,\text{kJ mol}^{-1}$

calculate a value for the enthalpy of solution of $MgSO_4$.

State whether you expect this salt to be soluble or not, with reasons.

29 Considering the solubility of sodium chloride and magnesium chloride, which would be expected to have the more positive entropy change, and why?

Given the information that Na^+ ions are hydrated by an average of five water molecules, and Mg^{2+} by an average of fifteen water molecules, explain which of these has the more negative entropy change.

Name one other entropy change that contributes to the final outcome and explain why the solution of both salts is a spontaneous process.

30 Hydrazine (N_2H_4) reacts with oxygen according to the equation:

$$N_2H_4(g) + O_2(g) \rightarrow N_2(g) + 2H_2O(g)$$

Given the data:
Bond energy: $E(N—N) = +158\,\text{kJ mol}^{-1}$
$E(N—H) = +391\,\text{kJ mol}^{-1}$
$E(O=O) = +498\,\text{kJ mol}^{-1}$
$E(O—H) = +464\,\text{kJ mol}^{-1}$
$E(N\equiv N) = +945\,\text{kJ mol}^{-1}$

calculate a value for the enthalpy change for the reaction.

KINETICS

31 On the diagram below:

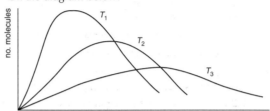

(i) What can you say about the temperatures T_1, T_2, T_3?

(ii) What can you say about the area under the curves?

(iii) How does the distribution of energy in a gas differ between T_1 and T_3?

(iv) Use the diagram to show why the rate of reaction will increase as the temperature increases.

32 A series of experiments measuring the rate of the following reaction with differing starting conditions is given in the table below:

$$2NO(g) + 2H_2(g) \rightarrow 2H_2O(g) + N_2O(g)$$

conc. NO(g)	conc. H_2(g)	rate
0.1	0.1	1
0.3	0.1	9
0.3	0.2	18

(i) Find the order of reaction with respect to both reactants.

(ii) Write the rate equation for the reaction.

(iii) Suggest the rate determining step for the reaction.

33 Suggest the best experimental technique(s) for following the rate for these reactions:

(i) $CH_3CO_2C_2H_5 + H_2O \rightarrow CH_3CO_2H + C_2H_5OH$

(ii) $H_2O_2(aq) + 2I^-(aq) + 2H^+(aq) \rightarrow 2H_2O + I_2(aq)$

(iii) $BrO_3^-(aq) + 5Br^-(aq) + 6H^+(aq) \rightarrow 3Br_2(aq) + 3H_2O$

34 Draw a reaction profile diagram to illustrate the effect of a catalyst on the rate of the reaction:
$A + B \rightarrow C$

CHEMICAL EQUILIBRIUM

35 The following equilibrium is involved in the industrial process for manufacture of sulphuric acid:

$$2SO_2(g) + O_2(g) \rightleftharpoons 2SO_3(g) \qquad \Delta H = -197\,kJ\,mol^{-1}$$

(i) Write the K_p expression, giving units.

Predict the effect on the equilibrium of

(ii) increasing pressure

(iii) increasing temperature

(iv) using a catalyst.

36 For the reaction:

$$CO(g) + Cl_2(g) \rightleftharpoons COCl_2(g)$$

(i) Give the equilibrium constant expression.

(ii) If the partial pressures of equilibrium are
$$p_{CO_2} = 2\,atm$$
$$p_{Cl_2} = 4\,atm$$
$$p_{COCl_2} = 48\,atm$$
calculate a value for K_p.

37 (i) Explain the term 'dynamic equilibrium' with reference to the reaction:

$$CH_3CO_2C_2H_5(l) + H_2O(l) \rightleftharpoons CH_3CO_2H(l) + C_2H_5OH(l)$$

(ii) Write the expression K_c for the above reaction.

(iii) Outline how you would get the results necessary to calculate K_c.

(iv) If 0.113 mol ester are mixed with 0.283 mol water, then at equilibrium 0.0569 mol ethanoic acid are produced. Calculate the value of K_c.

38 The radioactive isotope $^{214}_{83}Bi$ has a measured half life of 19.7 min.

(i) Write a rate equation for this radioactive decay.

(ii) If there were 10 g of starting material, how much of this isotope is left after 78.8 min?

39 (i) State the distribution law (partition coefficient).

(ii) If equal volumes of 2 mol dm^{-3} aqueous ammonia and trichloromethane are shaken together, calculate the concentration of ammonia in the organic layer, given that the distribution (partition) coefficient of ammonia between water and trichloromethane is 23.3 at room temperature.

40 Limestone is the product of continuous precipitation of calcium carbonate ($CaCO_3$) from sea water, over a long period of time.

(i) Write an equation for the solubility product K_{sp} of calcium carbonate.

(ii) Given the value of $K_{sp}(CaCO_3) = 5 \times 10^{-9}\,mol^2\,dm^{-6}$ and the information that the calcium ion concentration in sea water is approximately 0.01 mol dm^{-3}, calculate the maximum concentration of carbonate ions that can be present in the sea before precipitation.

ACIDS AND BASES

41 Define the term 'pH'.

Assuming complete dissociation, give the pH of:

0.1 mol dm^{-3} KOH(aq)
0.02 mol dm^{-3} HCl(aq)

42 How does the Brønsted–Lowry definition of acidity differ from that of Lewis?

43 Carbon dioxide dissolves in water to give a weak acid, often referred to a carbonic acid.

$$CO_2(aq) + H_2O(l) \rightleftharpoons H^+(aq) + HCO_3^-(aq)$$

K_a for this reaction is $4.5 \times 10^{-7}\,mol\,dm^{-3}$.

(i) Write down the K_a expression.

(ii) If HCO_3^- concentration is $2.5 \times 10^{-2}\,mol\,dm^{-3}$ and $CO_2(aq)$ concentration is $1.25 \times 10^{-3}\,mol\,dm^{-3}$ calculate the pH of carbonic acid.

44 Methanoic acid (HCOOH) is a weak acid, K_a value $1.6 \times 10^{-4}\,mol\,dm^{-3}$.

A buffer solution is prepared by mixing equal volumes of 0.1 mol dm^{-3} methanoic acid with 0.1 mol dm^{-3} sodium methanoate solution.

(i) Calculate the pH of this buffer solution.

(ii) Outline how it behaves as a buffer.

(iii) How would you change the solutions if you wanted a buffer of a slightly different pH value (say ±0.2 pH units)?

(iv) How would you change the solutions if you needed a buffer of significantly different pH value (say ±2 pH units)?

(v) If you were to attempt to perform a titration of the methanoic acid solution with 0.1 mol dm^{-3} NaOH(aq), what indicator would you select, and why?

REDOX REACTIONS

45 In the following equations give the oxidation number change for the elements in **bold**:

(i) $2\mathbf{Pb}S + 3O_2 \rightarrow 2PbO + 2SO_2$

(ii) $2\mathbf{Sb} + 2NaOH + 2H_2O \rightarrow 2NaSbO_2 + 3H_2$

(iii) $3Fe^{2+}(aq) + [\mathbf{Au}Cl_4]^-(aq) \rightarrow Au(s) + 3Fe^{3+}(aq) + 4Cl^-(aq)$

(iv) $2IO_3(aq) + 5\mathbf{S}O_2(aq) + 4H_2O \rightarrow I_2(aq) + 8H^+(aq) + 5SO_4^{2-}(aq)$

In each case state whether the change represents an oxidation or a reduction.

46 What is meant by the phrase 'the standard electrode potential of the Zn^{2+}/Zn system is $-0.76\,V$'?

47 The amount of iron in a sample can be estimated by dissolving the sample in acid to give $Fe^{2+}(aq)$ and then titrating samples of the solution with potassium manganate(VII) solution.

(i) Give half equations for all redox processes mentioned, and give a balanced equation for the filtration reaction.

(ii) Use the equation to find the concentration of the $Fe^{2+}(aq)$ solution, given that a $25.0\,cm^3$ sample required $23.8\,cm^3$ of $0.02\,mol\,dm^{-3}$ $MnO_4^-(aq)$ for completion.

48 Look at the following standard electrode potentials:

$Zn^{2+}(aq) + 2e^- \rightleftharpoons Zn(s)$ $\quad E^{\ominus} = -0.76\,V$

$Fe^{3+}(aq) + e^- \rightleftharpoons Fe^{2+}(aq)$ $\quad E^{\ominus} = +0.77\,V$

$Cr_2O_7^{2-}(aq) + 14H^+(aq) + 6e^- \rightleftharpoons 2Cr^{3+}(aq) + 7H_2O$ $\quad E^{\ominus} = +1.33\,V$

$Cl_2(aq) + 2e^- \rightleftharpoons 2Cl^-(aq)$ $\quad E^{\ominus} = +1.36\,V$

$MnO_4^-(aq) + 8H^+(aq) + 5e^- \rightleftharpoons Mn^{2+}(aq) + 4H_2O$ $\quad E^{\ominus} = +1.51\,V$

Then,

(i) Predict the outcome of adding $Fe^{2+}(aq)$ to acidified $Cr_2O_7^{2-}(aq)$.

(ii) What would be the e.m.f. of a cell formed by the above reagents, and what would the polarity be?

(iii) Give a balanced equation for any reaction.

(iv) Show by diagram how you could verify the e.m.f. value predicted.

(v) Why can dilute hydrochloric acid be used to acidify $Cr_2O_7^{2-}(aq)$ but not $MnO_4^-(aq)$?

INORGANIC CHEMISTRY

49 Explain what you understand by the term 'first ionisation energy of sodium'.

The values for the 1st and 2nd ionisation energies for sodium are $496\,kJ\,mol^{-1}$ and $4563\,kJ\,mol^{-1}$ respectively.

(i) Why is there such a large difference between the values of 1st and 2nd ionisation energy?

(ii) Predict the range of values for potassium and for magnesium, giving reasons.

50 Give the name of an element in Period 3 (Na to Ar) which:

(i) forms an oxide of type X_2O_3 only.

(ii) forms oxides of type X_2O and X_2O_7.

From the oxides of Period 3 elements, give the formula of:

(iii) one oxide with a giant ionic lattice.

(iv) one oxide with a macromolecular stucture.

Write an equation for P(III) oxide reacting with water, and suggest the likely pH value.

51 A student burnt magnesium in air to form MgO, magnesium oxide.

(i) What observations would the student have made?

(ii) Give an equation, with state symbols, for the reaction on adding water in which it partially dissolved.

The student suspected that magnesium hydroxide solution was formed.

(iii) Suggest a test by which this could have been confirmed.

(iv) Magnesium hydroxide is used as a remedy for indigestion. Suggest a reason for this.

(v) How successful would the student be in attempting to obtain magnesium oxide by heating magnesium carbonate?

52 Pure samples of hydrogen halides, HCl, HBr, HI are prepared by adding a suitable acid to corresponding potassium halides.

(i) Give the name or formula of an acid which can be used to prepare all three hydrogen halides.

(ii) A test tube containing hydrogen chloride gas is inverted in water. Describe and explain what you would see.

(iii) When hot wire is plunged into tubes of the three hydrogen halides, only one undergoes a change. What is the observed change? To the tube are added a few drops of sodium thiosulphate solution. What further change is observed?

(iv) Give an equation for both observed changes.

53 (i) Give the electron configuration for a copper atom.

(ii) Explain why copper is a good conductor of electricity.

54 A white copper compound A dissolved in water to give a blue solution of a cationic complexion B, and an anion C.

On addition of barium chloride solution to a sample of the blue solution a white precipitate, D, was obtained.

To a further sample of the blue solution, aqueous ammonia is added dropwise to yield a pale blue precipitate, E, and an further addition, a deep solution containing a complex cation F. Addition of EDTA solution to F gives a paler blue solution containing G. Identify A to G, and give their formulae. Explain the reactions using, as appropriate, stability constants.

55 Vanadium is a transition element.

(i) Give two characteristic properties of the transition elements, other than the ability to form coloured ions.

(ii) Outline the reason for colour amongst transition element ions.

(iii) Acidified ammonium vanadate contains the VO_2^+ ion and is yellow. In the presence of zinc the colour changes to green, then blue, green again, and finally violet. Give formulae for ions formed, and give an explanation for these changes.

REACTION MECHANISMS

56 Potassium hydroxide can be used in substitution reactions with halogenoalkanes. 2-bromo-2-methyl propane is known to react with hydroxide ions via an S_N1 mechanism. Outline the mechanism for this reaction, explaining the role of the hydroxide ions.

57 Outline the mechanism by which a Friedel–Crafts catalyst allows electrophilic substitution of benzene.

58 Butanone reacts with hydrogen cyanide in the presence of potassium cyanide.

  (i) Describe, with the aid of curly arrows, the mechanism of this reaction.

  (ii) What type of reaction is this?

FUNCTIONAL GROUP CHEMISTRY

59 Write equations, stating briefly the necessary conditions, for

  (i) converting propan-1-ol to propanoic acid.

  (ii) converting propanoic acid to propanoyl chloride.

  (iii) Either product can be used to prepare an ester with ethanol. Outline the differences in practical procedures.

  (iv) How might propan-1-ol be prepared through use of a Grignard reagent?

     Reaction sequence (i) above proceeds in two stages.

  (v) Identify the 'intermediate' product.

     Describe how it reacts with:

  (vi) sodium tetrahydridoborate (III) $NaBH_4$

  (vii) Benedict's solution.

60 In the following reactions scheme:

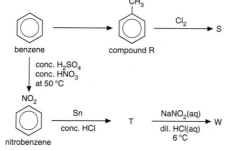

  (i) Identify compounds S, T, and W.

  (ii) Give a mechanism for the nitration reaction.

  (iii) Identify the type of reaction in forming T.

  (iv) Give one use of the product W.

  (v) Compound R will react with potassium manganate(VII) to yield an insoluble white precipitate. Identify the product.

  (vi) If the temperature in the nitration reaction is allowed to rise a different product with three isomeric forms results. Give structural formulae and names for the three isomers.

61 (i) Outline the stages in the laboratory preparation of bromobutane from butan-1-ol.

  (ii) Give the structural (graphical/diplayed) formulae for the products of the reaction of bromobutane with:

    ammonia, aqueous potassium hydroxide

  (iii) If the potassium hydroxide solvent is ethanol plus water, then the reaction with bromobutane is different. Give structural formulae, and name the products of this reaction.

62 (i) Draw structural (graphical/displayed) formulae of the isomers of amino acids with the molecular formula $C_4H_9NO_2$.

  (ii) Take one of the amino acid isomers and describe how it will react with dilute hydrochloric acid, and aqueous sodium hydroxide.

    Molecules of the amino acid alanine $CH_3CH(NH_2)COOH$ will react together. Use a structural diagram to show the product of two such molecules reacting together. Put a ring around, and name the linkage.

63 Look at the scheme below and answer the questions.

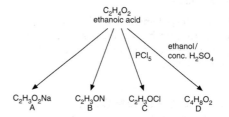

  (i) Name and give the structural formula of D.

  (ii) What type of reaction is the reversal of this (i.e. D → A)?

  (iii) What are the products of a reaction between A and C?

  (iv) B is an amide. What reagent and conditions are required to convert ethanoic acid to B?

  (v) B can be dehydrated to a further product E C_2H_3N. Give a dot-and-cross diagram for E.

  (vi) Write balanced equations for the conversions of ethanoic acid to A and D.

64 Salicylic acid is used as a precursor in the preparation of aspirin.

salicylic acid

  (i) Give the systematic name for salicylic acid.

  (ii) Identify the two functional groups in salicylic acid.

  (iii) How does salicylic acid react with methanol and with ethanoic acid?

  (iv) What would you expect to see if salicylic acid were added to neutral iron(III) chloride solution?

65 The skeletal formula for a particular hydrocarbon is written:

(i) Give the compound's full structural formula, and its systematic name.

(ii) What will be observed on shaking the compound with bromine water?

(iii) Write an equation for the reaction.

(iv) What isomers are possible if the compound is reacted with hydrogen bromide?

POLYMERS

66 Give an example, and use it to describe the formation, of:

(i) a synthetic polymer through addition polymerisation

(ii) a synthetic polymer through condensation polymerisation

(iii) a natural polymer through condensation polymerisation.

67 Why does the polymerisation of ethan-1,2-diol with benzene-1,4-dicarboxylic acid give a polymer with superior properties to that using benzene-1,2-dicarboxylic acid?

APPLICATIONS OF CHEMISTRY

(NB Much of the material in these pages addresses topics that may well be optional within a particular syllabus. You should check that sections are relevant to your particular course before attempting the revision questions.)

68 Outline what is meant by the terms 'primary structure' and 'secondary structure' of a protein.

69 How might a protein sample be prepared for thin layer chromatography analysis of its component amino acids?

70 What are the essential stages in the synthesis of proteins in the ribosomes?

71 What are the essential features of the mode of formation of:

(i) hydrothermal deposits

(ii) placer ore deposits?

72 Outline the need for catalytic converters on vehicle exhaust systems.

73 Give, with details, four ways in which the properties of a polymer might be modified.

74 What are the major factors which influence the choice of site for an oil refinery?

75 In considering the economics of the refining of crude petroleum (not its production), what is meant by (i) fixed costs (ii) variable costs?

76 Outline the photochemical reactions leading to the formation of a steady concentration of ozone in the stratosphere. How are CFCs affecting this?

77 Describe how a photocolorimeter can be used in a laboratory investigation of a coloured complex ion.

78 Draw a flowsheet for the various stages in the laboratory extraction of copper from a copper-containing ore.

79 List three different problems that can arise by a release of sulphur dioxide (SO_2) into the atmosphere.

80 Account for the fact that a green solution of chlorophyll looks red if viewed at right angles to a light source.

81 Why might rapid cooling of molten metal enhance its strength?

ANSWERS TO NUMERICAL QUESTIONS

1   (i) 0.018 moles (ii) 0.009 moles

2   (i) 0.020 moles (ii) 67.0%

3   (i) 1.36 g (ii) 97%

4   59.1

6 7.24×10^{14} Hz

12   207.2

24 $-665.8 \, \text{kJ mol}^{-1}$

25 $+844 \, \text{kJ mol}^{-1}$

26 $391 \, \text{kJ mol}^{-1}$

27 $-450.8 \, \text{kJ mol}^{-1}$

28 $-61 \, \text{kJ mol}^{-1}$

30 $-481 \, \text{kJ mol}^{-1}$

36 $K_p = 6 \, \text{atm}^{-1}$

37   (iv) $K_c = 0.256$

38   (ii) 0.625 g

39   (ii) 0.0827

40   (ii) $5 \times 10^{-7} \, \text{mol dm}^{-3}$

41   13, 1.7

43   (ii) 7.7

44   (i) 3.8

47   (ii) $0.095 \, \text{mol dm}^{-3}$

INDEX

Entries in **bold** type indicate main topic entries.